D0248901

A Rough Ride to the Future

JAMES LOVELOCK

A Rough Ride to the Future

The Overlook Press
New York, NY

This edition first published in hardcover in the United States in 2015 by
The Overlook Press, Peter Mayer Publishers, Inc.
141 Wooster Street
New York, NY 10012
www.overlookpress.com

For bulk and special sales, please contact sales@overlookny.com'
or write us at the above address

Copyright © James Lovelock, 2014

All Rights Reserved. No part of this publication may be reproduced or
transmitted in any form or by any means, electronic or mechanical,
including photocopy, recording, or any information storage and
retrieval system now known or to be invented without permission in
writing from the publisher, except by a reviewer who wishes to quote
brief passages in connection with a review written for
inclusion in a magazine, newspaper, or broadcast.

Cataloging-in-Publication Data is available from the Library of Congress

Manufactured in the United States of America

ISBN 978-1-4683-1046-7

1   3   5   7   9   10   8   6   4   2

I dedicate this book to my beloved wife Sandy

The planet has finally grown its own nervous system: us.

Daniel C. Dennett, *Freedom Evolves*

Contents

List of Illustrations

Acknowledgements

For many reasons this has been a difficult book to write, and I thank my friends for their unstinting support and for their patience. I am truly grateful to Richard Betts, John Gray, John Gribbin, Stephan Harding, Chris Rapley, Elaine Steel, Sir Crispin Tickell, Michael Whitfield and Dave Wilkinson. Especially, I am grateful to Stuart Proffitt, without whose skilled editing my book would be like a jigsaw puzzle waiting to have its separate pieces joined.

Introduction

This is not a book about climate change and what we should be doing to improve our carbon footprints – climate change comes into it, and the recent storms and inundations here in the United Kingdom and the cold breath of the polar vortex in North America remind us of that. What I am excited about, and write about in this book, is the extraordinary event that happened around 300 years ago, which put the world into flight to a destination where everything we now know about ourselves, the Earth and the universe will be different.

Life has flourished on Earth for billions of years because it discovered how to harvest the energy of sunlight and use it to reproduce and evolve. Among its successors were those, including us, that could recycle the carbon and oxygen of these early photosynthesizers. Like a hermit crab that makes its home in the dead shell of a whelk, life occupied, used and integrated itself with the material Earth and made it a living planet. A mere million or so years ago we emerged from the evolution of the primates as the first animal to harvest information, which is the truly fundamental essence of the universe; then, in less than a blink of the eye, 300 years ago we found ourselves at the beginning of a massive inflation of information harvesting.

As a consequence I find myself now in an extraordinary position for a scientist. In this book I argue that we humans may have reached the stage where we are one of the truly important species of the Earth's history and at least as significant as the photosynthesizers who eons ago invented the intricate process for harvesting sunlight for food and energy. In a way I am counter-intuitively supporting Bishop Wilberforce in his debate with T. H. Huxley at Oxford in 1860 on Charles Darwin's theory of evolution. The crux of their argument lay in the question: Are we humans merely another animal, or are we so important as to justify calling ourselves God's chosen species?

I see us as crucially important because we are the first species since life began over 3 billion years ago to harvest information massively and then use it to change the Earth. Of course, I do not really know whether or not we are, as the Bishop argued, God's chosen species. But I have little doubt about our irreplaceable value to the Earth, to Gaia. We are especially important now because if our form of life, from the smallest bacterium to the largest whale, was wholly destroyed from the Earth, life could never restart on the barren planet that would occupy our present position in the solar system. Preserving life on earth in some form is the challenge for, and responsibility of, humankind now.

A Rough Ride to the Future

I

What This Book Is About

Larvae and caterpillars are not the most attractive form of insect but they have that wonderful potential for changing themselves into peacocks, red admirals or painted ladies. But who grieves the passing of the caterpillar when it pupates in readiness for emergence as the butterfly? It just might be that the Earth is like a larva and could soon morph into a new and more exciting live planet. Should we be grieving for our old and familiar planetary home? Or filled with wonder at what the new forms of life will be? Or even full of joy about what we see as this imminent and timely change? More probably, I see us scared and confused, like a colony of red ants exposed when we lift the garden slab that is the lid of their nest.

We are right to be scared by the manic growth of industry, by climate change, by hunger in the face of an ever-growing population, and by the way the fitful wind of change blows over the market tables of economics. So what happened that can explain the onset of these disturbing events?

From the 1960s awareness grew that pollution might be more than a merely local problem, like smog in Los Angeles and London. The awareness became palpable when Rachel Carson published her seminal book *Silent Spring* in 1962. She is rightly remembered as the woman who inspired the start of the modern Green movement, and the public recognition that there is a dark as well as a beneficial side to chemical industry. A few scientists had been aware since the nineteenth century that the slow accumulation of carbon dioxide in the atmosphere could at some future date cause warming, but the distant problem for humanity of such global warming was almost ignored. It needed Carson's intervention to make us realize that our industrial lifestyle could affect everything alive, but what instilled the first frisson of fear was the discovery by Mario Molina and Sherwood Rowland that the otherwise harmless household chemicals, the chloro-fluorocarbons or CFCs used in spray cans and in our refrigerators, could by their presence in the atmosphere catalyse the destruction of the ozone layer and so be a threat to all life on the surface of the Earth.

This possibility that our emissions to the atmosphere represented a global threat much more serious than local pollution by smog, became the great concern in the 1970s and 1980s, and culminated in the Montreal Protocol (1989) banning the emission of CFC gases into the atmosphere. We did not then realize how potentially dangerous was climate change caused by fossil fuel combustion: the danger from the CFC emissions was by comparison a small problem and one that we have successfully dealt with.

Now in 2014 we are well aware of the threats that loom but there is little indication anywhere of a coherent, sensible response to our predicament, as there was with the banning of chloro- and bromo-fluorocarbon emissions. We are still behaving like those ants whose nest we disturbed in the first paragraph.

The pollution problems that come from chemical industrial poisons are relatively easy to solve. This is because chemical industry has evolved and is now mainly run by an intelligent and usually responsible technocracy and not by the bullish males of the nineteenth century that figure so largely in fiction and in the minds of politically active students. Indeed we may be hampered in our attempts to solve the large problems by the absurdly zealous application of health and safety laws.

The control of the emission of carbon dioxide and other greenhouse gases is far from easy. To start with, nearly 30 per cent of the total emissions comes from food production and consumption. This includes everything from the transport of food, fertilizers and the machinery of farmers to that part of the infrastructure of civilization that includes the storage, marketing, consumption and disposal of wastes. Most of the 70 per cent of remaining CO_2 emissions comes from industry and the transport of our goods around the world, and from what we do every day: heat or cool our homes, use electricity, drive our cars and fly by plane. We also add our personal increment by breathing. Did you know that as you exhale your breath contains 40,000 parts per million (p.p.m.) of CO_2? This is 100 times as much as is in the air, and is comparable with the exhaust gas of your car – and there are over 7 billion of us doing it, together with our pets and livestock.

To reduce CO_2 emissions effectively in the face of an ever-growing number of people is probably a task beyond the power of any government, democratic or dictatorial. This inconvenient truth emerged for me when I was in Paris at a meeting hosted by President Chirac on global change, and by chance found myself sitting next to Mario Molina, the Nobel laureate whose science first drew attention to the danger of CFC emissions. We turned to each other and agreed, 'How unfortunate that there are not just six large companies making CO_2 as there were with the CFCs.'

In the face of such overwhelming odds it seems naive of anyone to claim that they have a cure that will 'save the planet'. I have no remedy for the planetary diabetes that we and the Earth suffer, only a suspicion that we are overfed to obesity by our inability to stop guzzling from the industrial cornucopia.

Anyone who has had disease steal on them unawares and has lived a while with the sense that something was wrong, knows what great relief comes when told by a good physician just what ailed them; this can be true even when the news is bad. I will try in this book to tell you what we might do to reduce the adverse consequences of climate change and then suggest how we might learn to live with it and tell you why I think the final outcome for humanity may be better than we fear.

The best course of action may not be sustainable development but a sustainable retreat. On a battlefield when the odds against winning are seen as insuperable, a well-planned retreat is usually the best option. The need for such a retreat seems ever more confirmed as I write now, by a concatenation of potentially disastrous changes – to the climate, the economy, and the numbers of humans and their fellow-travelling species.

In the last three centuries we have changed our planet in a way reminiscent of one of the great changes that punctuated the evolution of the Earth since life's origin billions of years ago. Those of us who were in love with an earlier world where humanity and living things seemed to exist in a seemly harmony deplored the way we were busy destroying the world of Nature, the world of Rousseau, Gilbert White and the US naturalist Aldo Leopold. Even as a child I saw it as the greed-driven ruthless destruction of Nature. But what if we were wrong? What if it was no more than the constructive chaos that always attends the installation of a new infrastructure? Rarely does the building site of even the finest work of architecture look good.

Hubris has led us to believe that we began our journey into the future when civilization entered the Renaissance, a level that included genius of great quality among our scientists, artists, philosophers and leaders. We saw ourselves as so great that we were levitated above the commonplace. Looking back we saw this happening first in Italy in the fourteenth century, continuing until the seventeenth century in Europe, and then smoothly accelerating to the awful magnificence of now. Although it may seem to have happened like this, it did not. I think we moved into our new world because an ordinary man crossed its threshold in 1712.

The man was Thomas Newcomen, a blacksmith of Dartmouth in Devon, England. He devised and built a steam engine sufficiently powerful, at 4 kilowatts, and reliable to be able to pump flood water continuously from a coal mine at Dudley Castle in Staffordshire. The engine first worked in 1712 and the news of this small triumph of engineering soon spread, so that other mine owners were queuing to buy Newcomen steam engines. Important among the customers were tin and silver mine owners in Devon and Cornwall whose mining was also frequently hindered by flooding. The fuel preferred by the engines was coal, which was brought by sailing ships from places like Newcastle in the north of England, where news of the engines used to take flood water from mines stirred the interest of coal mine owners, and so steam power came to the north of England and Scotland. It was not long before inventors saw the possibility of making smaller, more efficient and compact engines that could be made mobile and could run on rails and be used to carry coal from the mines to the ships, or convey waste from the mines to dumps. In no time these steam engines were also powering the mills of manufacturers and pulling trains of carriages filled with passengers and the goods of commerce. Neither Newcomen nor anyone else at the time realized the significance of what had happened, but it was, nevertheless, the cause of our glory and our predicament. It may have changed the Earth forever. I think of it as the start of a new geological event, and the significant part of the newly named epoch of Earth science, the Anthropocene. A Newcomen engine is one of the first exhibits encountered in London's Science Museum today and is illustrated in Figure 1.

The term Anthropocene was coined by the ecologist Eugene F. Stoermer in the early 1980s as the name for the recent period of the Earth's history when mankind began to exert a noticeable effect on the living environment. The Nobel laureate physical chemist Paul Crutzen has sharpened and altered the definition of the term to include changes to the Earth's atmosphere and surface sufficiently great to be discernible by observers viewing the Earth from space. I feel honoured that Crutzen has taken my 1965 insight that the presence of life on a planet is revealed by chemical disequilibrium among

Figure 1. The Newcomen steam engine.

the gases of its atmosphere and used the even greater disequilibrium of the presence of unique gases like the CFCs to propose that the Earth has moved into a new epoch.

I like this new term, and agree with Crutzen and his Australian colleague Will Steffen that we need it to fasten in our minds the huge effect that the presence of technically enabled humans has had and continues to have on our planet. But I would urge them not to forget the similarly huge role of living organisms in shaping our planet. We do not yet know the details of life's origin, indeed we may never know them precisely, but among the many things needed was a generous supply of energy. For life it was sufficient that a continuous flow of sunlight gave warmth and light strong enough to make and break the chemical bonds of proto-life. I propose that the Anthropocene began when we first made a reproducible energy source of sufficient power to achieve an outcome that otherwise would have been effectively impossible. Previously we had man, horse, water and wind power but none of these provided a source of energy of sufficient power, economically and for continuous use over long periods. It is true that teams of horses could have been linked mechanically to achieve a similar result, but to have had those working twenty-four hours a day for many days is unlikely to have been convenient or economic. It was the invention of the first practical steam engine powered by burning coal that started the new epoch of the Earth.

The Anthropocene is not another word for the Industrial Revolution. No one knows when we first made and used tools on a scale large enough to constitute a tool-making industry. But by the time of the Romans industry was well established and has persisted ever since. We all know that the world we inhabit now is profoundly different from what it was before the eighteenth century. But it is qualitatively different. For once in history the transition to this new epoch, the Anthropocene has a clear and unequivocal date, 1712, the year that the Newcomen engine was built and did useful work.

The difference between the industrial epoch and the Anthropocene is profound and we are only now beginning to see and feel its awesome presence. I will try to explain in Chapter 3 how the emergence of the Anthropocene is, like many physical phenomena, to be connected with the flow of energy. I find it intriguing that the energy flow needed to keep Thomas Newcomen's steam engine going is comparable with that the Earth received from the Sun when life began more than 3.5 billion years ago. It was about one kilowatt per square metre. The easiest and most readily available artificial source of this intensity was the burning of the fossil fuel coal.

I will follow Crutzen's example and appropriate Stoermer's word but with an even sharper definition. This is badly needed for otherwise this clear and useful term is in danger of losing resolution in the noisy background of vague academic niceties and amorphous thoughts about ecological sin. Whereas in fact the emergence of this crucial period may change the Earth and its future as much as did the origin of life on Earth more than 3 billion years ago.

So who was Thomas Newcomen? He was the village blacksmith of Dartmouth in Devon and had served an apprenticeship as an iron-monger, but although he was literate and numerate and was a lay preacher in Dartmouth, he was not academically qualified. He was a good inventor and made an engine that worked using his skill as a smith. England was favoured because of the availability of coal; some of it was even on the surface. The crucial point is that burning coal produces ten times more energy than burning the same mass of wood, the alternative fuel. The successful use of Newcomen's engine, I hypothesize, was the first occasion in the Earth's history that a sustained continuous source of energy above a critical threshold flux of approximately 1 kW per square metre was used purposefully, successfully and economically for periods as long as days. More than this, thousands of copies of his engine were made by other inventors until James Watt improved it so much that a new generation of efficient, convenient and practical steam engines appeared. I see this as the start of a new evolutionary process that soon became one million times faster than Darwinian evolution by natural selection, and it was one that proceeded in parallel. If the annual rate of invention is extrapolated backwards, its exponential slope flattens in the eighteenth century, which is another reason why I suggest that we take 1712 as the start date for the Anthropocene.

I think it important to repeat that the new epoch is not, as commonly assumed, loosely connected with the presence of industry or of pollution, which could have been at any time from the Roman civilization until Eugene Stoermer first noticed in the 1980s that the once pristine waters of the Great Lakes were slightly contaminated by the chemical wastes of industry. It had to wait until there was a unique artificial source of heat energy of sufficient intensity and duration to sustain useful work. Most important of all, this new source had to be readily and economically reproducible and have available an unlimited source of fuel. These criteria and the pressing need of mine owners with flooded mines are what set the start of the Anthropocene with the steam engine in 1712.

For consistency, this book will use word Anthropocene frequently because there is no alternative that is precise enough. The term 'industrial revolution' is so imprecise that by comparison it seems to have lost meaning, except as a political slogan.

An insight that emerges from these thoughts is that the advanced technology of today owes at least as much to the craftsmanship of a talented engineer as it does to science. The ability of individual brains to think rationally and analytically came much earlier, at least as far back as the civilization of Ancient Greece, and the sum of our knowledge from those early thoughts, observations and experiments became what we know as science. It flowered within European civilization as the Renaissance, from the fourteenth to the seventeenth centuries, but it was not science that started the Anthropocene, or science alone that made the modern world.

As a scientist I should feel chastened to discover that only a small part of this new epoch came about as a consequence of science, or the rational thoughts of a great philosopher. But I still feel as much devoted to science as I imagine a saintly priest is to God. My creed is simply to accept that nothing is certain, but always a matter of chance. Although to be confronted with the concept that irrational invention, not science, is what moves us forward should have been a painful surprise, in reality it was not. Good scientists have come to realize that the greatest revelation so far is that of natural selection, and it operates across the whole of science, not biology alone. Jacques Monod described the evolution of life as a consequence of chance and necessity. It is necessity, the parent of invention, that links together

Darwin's great thought with all of what I accept as science. Perhaps it is what makes me as much an inventor as a scientist.

The eminent biologist William Hamilton taught that a peacock with a flawless tail reveals to the peahen a male fit to be the father of her progeny, so we might consider the possibility that the human conscious mind was selected for its capacity to tell entertaining stories and reveal to a woman someone lively and fit enough to be the father of her children. The peacock invests a great deal of evolutionary and metabolic effort in his extraordinary tail, and so do we in our brains. Could men's skill at 'chatting up the birds' have been selected as a measure of fitness? While this trait may have played a part in the selection of our brains, there is clearly much more to it than this alone. Peahens have small tails, but men and women have close to the same size brains. Could the hidden and often unappreciated parts of our brains be those that have been selected, accounting for its large size? I have in mind the qualities that enable a child to become an adult skilled in a vast range of capacities from the visual arts through music, writing in all its forms and, most important, everything we think of as creative, especially hands-on as well as intellectual creations. In technical terms, the human brain is a computing device that can extrude professional-quality peripherals, such as the hands and arms of a concert violinist, a master surgeon or a blacksmith like Newcomen. It can do all these things as well as think and talk or write about them.

I first heard about the brains of whales in the 1950s when I was a staff member of the National Institute for Medical Research in London, researching blood lipids and their significance in heart disease. A colleague, Griff Pugh, who had travelled across Greenland as part of a group of men, hauling sledges across that icy place, had brought back blood samples for me to analyse. The sledge party had been consuming more than 7,000 calories of food daily (about twice the intake of a manual labourer), and supplementing their meals with milk chocolate spread with butter. Yet analysis of their blood showed a near ideally healthy mix of lipids. This was not good news for those who believe that an excess consumption of saturated fats is unconditionally unhealthy, but that is another issue. My physiologist colleague had also examined the brains of large whales when travelling on a whaling ship. From him I learnt that whale brains contained several times as many neurons as human brains and might therefore be thought to have that much greater capacity for thought. I have wondered ever since about their thoughts and the only, to me, useful conclusion was that great brains alone are not enough to dominate the Earth as we do.

Our mistake when we started the Anthropocene was to fail to notice that inadvertently we were the catalyst of the reaction between the carbon of coal and the oxygen of the air. I like to think of us as like the bacteria *Bacillus subtilis*, which feed on the new damp grass of a haystack; as they eat, the heat from their metabolism adds to that from the inorganic aerial oxidation of the grass. Deep in the stack and well insulated by the hay, these two heat sources raise the temperature and positive feedback accelerates the rise in temperature until the grass is hot enough to burst into flame, an example of so-called spontaneous combustion.

With hindsight, we see industrial growth as a human triumph that gave us the key to a technological paradise where we could live out full, rich lives and reach a deeper understanding of ourselves, life and the universe. To an extent this promise has been fulfilled. But an utterly unbreakable law of the universe (the second law of thermodynamics) warns that nothing is free. Renewables such as water or wind energy may seem free but in truth they are second-hand sunlight. The warmth of the sun evaporates the water that falls as rain on the hills or heats the air enough to make it expand and rise like a hot-air balloon and so draw in the wind. Coal comes from trees that took in sunlight long ago and used it to grow. The coal is a tiny part of what was left behind in the soil when the trees died. Like the wind, coal is stored solar energy. The difference between them is the time taken for their renewal. It is not the cost in energy currency that matters to us; the cost lies in the side effects. CO_2 from coal causes a greenhouse effect. The huge subsidies needed for wind and water power impoverish our chance to spend on adaptation: think of the flood defences now needed in these islands.

The nineteenth-century physicists Rudolph Clausius and Josiah Willard Gibbs formulated the second law of thermodynamics that states that the spontaneous flow of energy in the universe is always downwards towards an ultimate sink. The energy all comes from that left over after the Big Bang 13.8 billion years ago and is slowly used as the stars burn and die and as the universe expands, and also contracts locally to form black holes.

The second law is too often used as a metaphor for gloom and death. It certainly does imply that life, the universe and everything is destined to die eventually but, as with a 99-year lease on your home, there is ample time to enjoy life before the lease terminates. I am a cheerful man and have long thought that the second law could be a source of a much happier metaphor. When we use a flashlight it allows us to take the energy stored in its battery to make the light that shows the way ahead. In a similar way all of the energy we use on Earth comes from the Sun, which we could regard as a long-life battery; eventually it will have none left to deliver. But we do not fear that the Sun will not shine tomorrow morning.

You can choose to think that the dark side of the Anthropocene was entirely our fault and we are guilty of a great sin, or that it all happened because mine owners were greedy capitalists out for personal gain and did not hesitate to plunder the Earth's riches (and if the mines had been run by a miners' cooperative venture, it would all have been done for the people's benefit).Whatever way you see it, the truth is that humans were merely innocent agents who did not realize that they were catalysing a strongly positive feedback process that would grow exponentially until much of the easily available fuel had been consumed; and that is where we are now.

This book attempts to explain the consequences of this extraordinarily powerful combination of events – not just the climate change, or economic growth, or even the explosion of human population and invention. Its theme is the effect of all of it on the great Earth system, Gaia, and on our futures.

The release of the Earth's stored energy was an event waiting to happen, and it took only the skills and determination of a single inventor to make this energy available to humanity. Our mistake, if it was a mistake, was the failure to notice the consequences of this new and powerful source of energy on the pace of invention, the growth of population, the climate, the means of exchange and monetary inflation. Taken together these changes now imply that we are experiencing an entirely new form of evolution. This is the significance of this new epoch, the Anthropocene, which otherwise is no more than another division of the Earth's history into periods convenient for teaching geology. And importantly, the new evolutionary process is something quite different from all that went before it, and may mark the end of the primacy of evolution by natural selection which has carried us and the Earth for the past 3 billion years at its slow, unhurried pace. The impact of this new form of evolution affects all species, the planetary environment and, of course, human civilization. As we shall see, some parts of it are evolving one million times faster than happened in the pre-Anthropocene world of evolution by natural selection.

Another subject that I will touch on as we proceed is the difference between the scholastic, rational side of science and the instinctive and intuitive approach that inventors and creative scientists use, although convention rarely allows them to say so. As we shall see, the scientist Faraday and the blacksmith Newcomen were brothers under the skin.

One might say that if Gaia's goal is to keep the Earth always habitable, it was surely irresponsible to make this huge store of combustible material and leave it in an oxygen-rich atmosphere. In Gaia's defence I would point out that her practice of burying 0.1 per cent of the carbon photosynthesized by plants was essential to sustain the 21 per cent abundance of oxygen in the air (H. D. Holland, *The Chemical Evolution of the Atmosphere and the Oceans*, 1984). We regard 21 per cent as the natural abundance of oxygen in the world we were born and live in. It isn't; this level is merely the steady-state equilibrium abundance of the contemporary Earth system. We are mostly, especially our governments, scientifically illiterate, so we rarely wonder why the air has 21 per cent of oxygen, even though without this much oxygen we and other animals could hardly move and certainly would not be able to fly or think; indeed, we would not have evolved as we have done. If oxygen were only 13 per cent abundant there would be few if any fires, but at 25 per cent they would be furious and uncontrollable. In fact, what we think of as our coal and oil reserves represent only a tiny proportion of the total carbon buried, almost all of it present in dilute form and doing no more than making the sedimentary rocks darker than otherwise they might be. (We are fortunate that there is no inbuilt health and safety system in Gaia, otherwise the dangers of fires would have led to the banning of its production.) Evolution is a risky business and we are lucky that it is blind and its mistakes are corrected by natural selection. Fires can be devastating locally, but without a sustained oxygen-rich atmosphere the Earth would now be a dry and barren desert like Mars. This is why oxygen regulation is so important. Oxygen molecules and atoms behave like guards that patrol the frontier between space and the edge of the atmosphere, and they capture the fast-moving hydrogen atoms that otherwise would escape the Earth's gravity. The fugitive hydrogen is taken back tightly bound in the water molecules from which it escaped.

Were you taught, or can you explain rationally, exactly why you can stand, or keep upright on a moving bicycle? Have you ever thought about the meaning of consciousness? Or do you merely take these things for granted? Questions like these are inevitable whenever we try to understand anything that self-regulates, such as a spinning top, a bacterium, a cat, a tree or ourselves. These all share in common the ability to sustain a steady state by being dynamic and self-regulating; this I see as an important property of all living things and something we share with Gaia.

Darwinian evolution simply says that the successors are those organisms that leave the most progeny. We have acquired the ability to make artefacts, and their further evolution is under our control: we can choose only the best and then make more of them. But how long will it be before they are much faster and more efficient than we are, and then begin to select themselves?

If the Anthropocene were the defining factor that led to modern civilization, then we have to ask, how much of it can be attributed to the inspiration of talented individuals and their flowering during and after the Renaissance? More likely, I think, the burgeoning progress we see around us, good and bad, may have come from a simpler and cruder source: that is, the work of rude mechanicals who worked blindly like Wagner's Nibelungen, who made their Ring with no thought for the consequences.

No one knew enough in the eighteenth and nineteenth centuries to realize how large a change we were making. Indeed it was not until near the end of the twentieth century that we began to notice that the climate and chemistry of the Earth had altered, and then considered that the Earth might be a self-regulating system kept stable and habitable by feedbacks between life and its non-living infrastructure. Until recently we thought, and to an extent still tend to think, of our planet as a ball of rock covered by a thin film of air and water, with abundant life as something separate. Yet life is not separate but is an integral part of the real Earth system whose power to self-regulate might be compared with that of a modern airliner that can fly and land itself should the pilot be disabled. Never underestimate the intelligence of the autopilot of a modern plane. It can take off and land despite the fickle weather, and navigate a fuel-efficient path to your destination. The pilot is there only because we do not yet trust computers. The ability of our artefacts to be, like the airliner, almost autarkic implies that they have a form of intelligence.

Martin Rees, a past President of the Royal Society, warned in his book *Our Final Century* (2003), of a world that embodied the experiences of the sorcerer's apprentice, a technological world entirely beyond our control. This has been in the minds of other prophets, especially Vernor Vinge and Ray Kurzweil. Until 2011 I tended to share the belief that progress was benign and that negative feedback from common sense, market forces and social instincts would sufficiently curb wild invention to keep us safe. But invention is proceeding not haphazardly, as we had assumed in the past, but exponentially. A part of this growth follows Moore's Law, first noticed by Carver Mead, a colleague of Gordon Moore, one of the founders of the microchip company Intel, who noted that the number of components on one square centimetre of computer chip doubled about every two years. This has happened now for nearly fifty years, and if there is a direct relationship between the number of components and the capacity of the computer chip, it implies an improvement of one thousand trillion (1015) times, a furiously fast rate of evolution – although there has been a less impressive improvement in the performance of our computers, and so far it has not led to the emergence of a cybernaut equivalent to an animal or a human. It is right to ask, what is the probability that it will, and if so, how long will it take?

Despite these awesome possibilities, I think that we may muddle through into a strange but still viable world. The Anthropocene is indeed a frightening prospect but intuition warns me that it may be wrong to assume that there is a direct relationship between the number of components in a computer and its performance. Support for this warning comes from the practical experience described in Chapter 9. At a guess this small piece of information suggests that improvement goes as the fourth root of the number of components. But it is no more than a guess: while I greatly doubt that there is a simple direct relationship between number of components and performance, the real connection is open for experimental measurement.

I am convinced that we are wrong to assume that the large problems that confront us are singular and to be treated separately. Climate change, population increase, economic growth and an ever-developing human civilization are all responses of the Earth system, and the driver of these changes is cheap and abundant energy which enables an exponentially growing rate of invention. We have tapped this source of energy since 1712 but there are signs that its availability is decreasing. Fortunately for us, in real world systems an impassioned and impetuous driver is not what determines the speed. Natural resistance, braking, or what systems people call negative feedback – all oppose the acceleration and confirm our experience that growth always eventually slows and halts. I am encouraged by the observations of climatologists, economists and demographers who report that the rapid growth they expected shows signs of failing to materialize. The slowing of the rate of growth is not much more than anecdotal. The global average temperature has not risen as expected, but this could be because we have not observed a long enough period for a confident assessment, and the same applies to the observation of population growth and economic decline. But if I am right in thinking that accelerated evolution is consequent upon the availability of cheap energy, then the slowing is consistent with the recent rise of carbon fuel costs.

Perhaps the decline in cheap energy, if it continues, will allow us and Gaia a chance to pause and catch our breath and then move into the future at a pace that gives time to appreciate the scenery. There are many consequences, good and bad, from our 300-year-long burst of invention, but there are no feasible ways to restore the climate to what it was, nor to reduce our numbers to those that so worried

Thomas Malthus at the end of the eighteenth century. Strangely, the greatest danger might be the invention of a simple, safe, inexpensive and unlimited supply of renewable energy, such as the one described in Chapter 7 of my first book *Gaia* (1979), for then we would be rushing to hell in a handcart powered by a damaging excess of renewable energy. If there is a message in this book it is that we are not yet sufficiently intelligent to control or regulate ourselves or the Earth.

Like it or not, we are part of Gaia, and like the citizens of a great nation we draw power from our membership. In common with all animals we have breathed in oxygen from plants and used it to recycle, as carbon dioxide, the food the plants provided. Now, through our intelligence, we have allowed our planet to become aware of its environment in space and not only to see its place in the cosmos, but also to grow aware of potential threats, such as that posed by an incoming planetesimal, one of the kind believed to have ended the reign of the dinosaurs. Because we are alive, in a rudimentary way the system has, through us, become sentient. Before this, life existed without knowing what it was, how old it was, or anything about its future. We are now travelling along a path that could lead us to become the citizens of a live, intelligent planet, which might in turn become a citizen of the galaxy. With such a future ahead of us how could we possibly be gloomy, or believe, as today's puritans keep telling us, that we are guilty of some great harm? We merely have to stop making mistakes, or better – because mistakes are inevitable – learn from them and keep our eyes on the path ahead.

For as long as I could think about it I have had an unconditional love for the natural world outside the city, and therefore thought that I was a 'green'. Urban and suburban green ideology talks of saving the planet, but it has mainly become another radical political movement, no longer concerned with the Earth, only with people. It is so different from my naive and simple thinking that it would be wrong to regard this book as 'green' in the modern sense. I find it sad that urban greens judge most of us guilty of causing adverse climate change, and do so without due process of moral and scientific evidence. There is no reason for us to feel guilty. To be sure we have made a mistake on a planet-sized scale, but out of ignorance not culpable negligence. We have not added CO_2 to the air just to make the Earth warmer as an act of geoengineering; we merely used the heat from burning wood and straw, first to cook, then for warmth and lastly as a source of energy. As soon as we discovered that coal and oil from the ground would burn, we used them instead of wood. We should have noticed that adding CO_2 and other greenhouse gases to the air would have consequences, some but not all of them unpleasant. None of this would have mattered if we had somehow kept our population to 1 billion or less.

If we could see the distant future of the Earth I think we might see it moving well beyond the Anthropocene, moving to a time when the dominant life form is no longer solely a wet carbon-based organic entity. It could instead be one that had evolved by the process of endosymbiosis discovered by my friend and colleague Lynn Margulis, so that a cooperative life form of wet organic chemical life and dry electronic life evolved in synchrony. This idea is discussed further in Chapter 10. Eventually, as the Earth grew hotter, and suitable for electronic life forms, it would be even more inhospitable to us than our world would have been to our Archaean ancestors. But these imaginary electronic life forms, based on semiconducting elements or compounds, might fill the body of a new form of life and take over from us the task of sustaining a self-regulating planet with an environment that would be sustained always at a habitable state for them.

Because we're full of pride and see humanity firmly established as the rulers of the solar system, we tend to think that nothing more powerful, more moral and more delightful or in any way better than we are could possibly come after us. Most of us find it difficult even to contemplate the possibility that we are fulfilling a role like that of the feathered egg-laying reptiles who were the predecessors of the birds. But the signs are there: we already talk, first in fiction but now in science, about artificial intelligence. There are many possible forms of life and evolutions of Gaia beyond the ones we know.

The notion of an artificial intelligence, a robot or a thinking computer that possesses the human capacity of empathy, is hard to take seriously, even as quality science fiction. Yet, as I will try to describe in the next chapter, we have been fully engaged in the project of changing the world into one suitable for such beings. I see nothing bad or wrong in this idea; indeed we might have cause for pride about the way we have changed the Earth's scenery in preparation for the next act of Darwin's evolutionary play.

2

The Lone Practice of Science

It was an article by the journalist-philosopher Jonah Lehrer in the *Wall Street Journal* in 2011 that made me think that intelligence, pure intellect, might not be all that it was cracked up to be. In his article, Lehrer argued that the days of the lone scientist were over and if the equals of the great individual scientists of the past – Galileo, Newton, Leibniz, Darwin and Einstein – appeared today they would find no place in the modern world of science. Science, he wrote, was now so complex and expensive that only governments and large corporations could afford to support it. Successful science, he seemed to imply, no longer came from the lone conquest of a scientific Everest. The modern world demanded the contest of hugely expensive teams in the science equivalent of an Olympic stadium.

My first instinctive thought was that this was dangerous nonsense. Great brains could function now as well as or better than in earlier times. But then I realized that he was at least partly right. When I started my practice as a lone scientist inventor in 1961 the bureaucratic restrictions were mild and easy to overcome, but now, more than fifty years later, they are formidable. In most nations of the developed world they rule out the greater and more interesting parts of hands-on science. True, it might be possible for a present-day Descartes, Einstein or Newton to think and use paper or a PC to record and expand their thoughts, but a Faraday or a Darwin would be buried in paperwork and obliged to spend their time solving problems concerning health and safety and political correctness, today's equivalent of the theocratic oppression of Galileo. In the world of corporate science there would be little time left for their singular and breath-taking ideas. More than that, the Internet has made the human world a monstrous village with an ever-growing population of nags, scolds and officious fools; soon, I fear, we face a life like that Japanese nightmare – the one that sees an outstanding brain as like a nail that stands out and is always hammered in.

For the past forty years I have worked alone in my own laboratory but as part of a rich life within a family and a village community. It is a mistake to regard a lone scientist as an unnatural or pathologically disabled person; I do not think that I was disabled or even lonely. What I mean by a lone scientist is one who is autarkic and does not need immersion in a think tank to excite ideas, which arise naturally through wondering. It would be easy for me to be a lone scientist within a good university department or scientific institute, provided that social, tribal and bureaucratic influences were minimal. You can easily distinguish lone from communal scientists by the authorship of their papers. The true loners write alone or with one, or rarely two, colleagues. I do not scorn the captains and colonels of science who lead successful teams, but we need the loners as well.

Everyone thinks they know what a scientist is and how she or he works. Figure 2 illustrates a serious laboratory where white-coated scientists conduct their experiments. Figure 3 illustrates my home laboratory as it was in 1981, where I was about equally engaged in invention as in science. You will note that I was not white-coated: I never felt the need to wear a uniform. I merely wore old clothes that could be discarded if contaminated.

My laboratory in no way resembled the work environment of a senior scientist in today's world. He or she works from an office and has a sizeable staff to perform practical experiments or plan expeditions to observe the Earth's near and far environment. More and more, the exciting and slightly dangerous experiments with chemicals, high voltages and radioactive substances are done by computer simulations. From my viewpoint, science lost its glamour about thirty years ago. No doubt the few surviving dinosaurs 60 million years ago felt the same about the safer mammalian world that was thrust upon them. Those in the arts know well the delights of hand and eye creativity and the true freedom it brings, but it is now so rarely found in science that I feel an urgent need to recommend at least the trial of the lone practice of science, if only because it is a way to break the disciplinary integuments that stifle scientists and science itself.

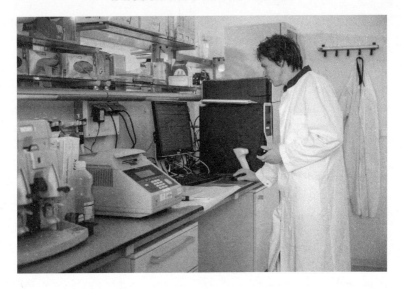

Figure 2. A modern laboratory.

Figure 3. My laboratory, 1981.

I have always, from childhood on, regarded science as a calling, a vocation, never as a career. For this reason I chose employment as a laboratory assistant in the late 1930s to learn the craftsmanship of science; the next twenty-three years I spent doing post-graduate medical research, almost all of it as a tenured staff member of the National Institute for Medical Research (NIMR). This was needed to round off my apprenticeship to science as a professional. From 1961 to 1964 I was employed as a research professor at Baylor College of Medicine in Texas, but during that time I fully developed my vocation as a lone practitioner. I even have the ultimate professional qualification: CChem, a chartered chemist of the Royal Institute of Chemistry. There is nothing quirky about this way of life, but it does differ from that of most professionals because the bulk of my income went towards the science I did rather than to improve the standard of living enjoyed by me and my wife. I aimed towards maximum economy and saw no point whatever in buying expensive equipment, because I knew that such apparatus (even when said to be the latest) was probably ten years out of date. I knew as an experienced inventor of scientific instruments that it takes many years to develop the working model of an idea into practical, saleable hardware. I could invent new equipment that was truly at the leading edge, so why did I need to buy what was already outdated?

I have never kept count of the many inventions I made but it must run into the hundreds. Most of them were trivial, such as a wax pencil that would write clearly on cold wet glassware straight from a refrigerator. It was published as one of my first letters to *Nature* in 1945. Although I am not the formal inventor with a patent I am fairly sure that I was the first to make a microwave oven powered by a one kilowatt magnetron. It was practical and was used almost daily for several months. It consisted of a metal cage similar in size to the one in a home kitchen, and was equipped with a timer. It was used to reanimate chilled small animals and to cook my lunch at the National Institute for Medical Research in the 1950s.

If there was something very complex that I could not easily make, like an electron microscope or a new form of mass spectrometer, I considered solving my problem another way, or sought a friend who could sell or donate spare time on his instrument, or could do the job for me. The one exception to this rule was the purchase of the best computer I could afford. It so increased my productivity that the cost was justified.

When I look back I am surprised by how often inventions stole into my brain when someone entered my room and asked, 'Can you think of a way to do ———?' An example easy to recall is the sudden appearance at the entrance of my lab in wartime London in 1943 of my boss, the physiologist R. B. Bourdillon. He said, 'Lovelock, can you make for me an instrument that will measure heat radiation accurately and record if the heat flux was enough to cause a first-, a second- or a third-degree burn on bare, exposed skin? I need it by ten tomorrow morning for an important meeting at the War Office.' It was then about 4 p.m.

He did not need to tell me that whatever device I invented, it had to be soldier-proof: that is, robust enough for use on a battlefield and capable of use after a few minutes' training. I knew immediately that no laboratory instrument with glassware or the primitive electronics of the 1940s would do. It had to be very simple indeed. Suddenly I realized that all that was needed was a sheet of paper that changed colour in clearly different ways for the three degrees of burn. Next morning I gave him a few sheets of paper that had been spray-painted lightly with a green paint. Exposure to radiant heat changed the colour from green to pink, scarlet and yellow for the three degrees of burn and it provided a permanent record of the exposure.

From the expression of the need to the product took about four hours of thought and experimental test in 1943. I suspect that posing the same need now to a civil service laboratory would provide several months' work for a team of scientists and technicians. The crux is need. In 1943 it was the solution of an urgent real-life problem; now in peacetime the need would be the sustenance of employment, expansion and preferment.

Another useful invention was a simple apparatus to lower the temperature of bull spermatozoa to -80°C, at which they could be stored in a viable condition for weeks or longer. But probably the most memorable of my inventions was the Electron Capture Detector (ECD), which has had a profound effect on our views on industrial poisons that either innocently or carelessly have contaminated the natural environment, and which was intimately involved in the consequences of Rachel Carson's revelations in her famous book *Silent Spring*. At the time it was published I was a science adviser to Victor Lord Rothschild, who was then the science coordinator for the Shell Oil Company. Shell also made DDT and other insecticides. But I had invented the ECD in 1957, and its exquisite sensitivity to DDT and other poisonous substances used by farmers fully confirmed Rachel Carson's claim that pesticides were everywhere in the world.

The ECD is so sensitive and so discriminating that it can detect as little as one-tenth of a picogram (10^{-13} gm) of DDT in air, water or food. At the time Rachel Carson wrote *Silent Spring*, the least quantity the chemists who advised her could detect was about one microgram, in other words their analytical methods were ten million times less sensitive than the ECD. Lawyers acting for the chemical industry could easily have argued that the evidence Carson presented was insufficient to justify a ban, but the ECD measurements allowed confident assertions that DDT was an environmental hazard. More than this, the ECD has the uncanny ability to ignore the harmless part of the environment and single out for attention only the poisons. Poisons like tetra methyl lead that used to be added to petrol to improve the performance of cars, the polychlorinated biphenyls (PCBs) that were ubiquitous in industry, the carcinogenic hydrocarbons that once shortened the lives of chimney sweeps and those whose work brought them in contact with partially-refined petroleum products: all of these poisons could for the first time easily be found in places where they could do harm.

By confirming Rachel Carson's findings the ECD transformed the ethos of the green movement; it was for chemicals what the Geiger counter was for radiation. Yet I do not think it could have been made as early as 1957 by any but a lone scientist, and this is why.

As the years have passed false fears of radiation have been amplified and have led to legislation that makes it difficult to make, use or sell any equipment that contains a radioactive source of the strength needed in an ECD. In fact, a typical ECD contains 10 to 20 millicuries of the isotope nickel$_{63}$ in the form of a thin layer of metallic nickel$_{63}$ plated on the inner wall of an inert nickel chamber, about one millilitre in volume. The radiation emitted by the nickel is in the form of beta particles – electrons – whose energy is so low that none penetrate the walls of the detector, but which are sufficiently powerful to release an ample supply of free electrons in the detector chamber. It takes rampant prejudice against nuclear radiation to think of a scenario that would make the ECD a radiation hazard. I would not expect any radiation pathology if such a detector had been in my trouser pocket for eighty years. Indeed it is far less dangerous than the wristwatches most of us used to wear, which were made luminous by the radium isotope they contained and emitted high-energy gamma radiation that penetrated our whole bodies. But in the USA it became a criminal offence to take an ECD across a state border – for example, from Missouri to Kansas. Fear of radiation has led to laws that trump all common sense.

Invention, science and war have always been tightly connected. Newton's laws of motion were born from the needs of the British Navy, and Darwin travelled on his amazing journey of discovery as a naturalist employed on the naval ship the *Beagle*. This relationship has continued ever since, and it may not be by accident that the best customer for my independent science and invention has been the British Ministry of Defence.

Humanity and science were offered a cornucopia of benefits from the accelerated inventions of the Second World War. Had we been less combative animals we could have used this new knowledge constructively. We could have made the observation of the Earth from space a priority, built satellites that viewed the land, the air and the oceans, and seen the looming dangers of global warming in time; instead we made space missiles. Such a statement sounds good when claimed with liberal hindsight. But if we think a little more deeply we have to ask, what university department or scientific institute anywhere would have bothered to spend as much as the Manhattan Project did on nuclear energy, or in peacetime thought of using rockets to put satellites in space. Such a hugely expensive project would have been described as 'Utter bilge'. These indeed were the words of Britain's Astronomer Royal in the 1950s when asked about artificial satellites. It needs the combativeness and the tribal anger of war to undertake such endeavours, and this is the crux of the global warming problem;

we do not yet regard it as serious as a major war. It may in fact be even more serious.

In the 1960s I worked for NASA at the Jet Propulsion Laboratory (JPL) in California and got to know the real rocket scientists, many of whom were not infinitely wise, nor did they all speak with Middle European accents and stand before a blackboard writing cadenzas of incomprehensible equations. Instead, those I met built those wonderful spacecraft that let us know for the first time the nature of our solar system and see the planets almost as sharply as if we were on them ourselves; most of them were Americans in early middle-age, including outstandingly competent engineers and astrophysicists. Surprisingly, to me, there was a sprinkling of older men who previously had been skilled in building intricate and beautiful things such as tiny yet perfectly working cars, railway engines, clocks and so on; a few of these rocket scientists and craftsmen were hands-on people with relatively few qualifications, but enormous expertise. Their models were real, solid, and worked. I was comfortable with engineers of this kind, people who understood thermodynamics and could also build exquisitely crafted model railway engines. I am proud that some of them came to my home in England and went with me to see the thirteenth-century mechanical clock that is on display in Salisbury Cathedral, then a few miles from my home. These were the men who built those early spacecraft *Pioneer* and *Mariner*, and sent them on their long journeys. *Pioneer* is still travelling and is now beyond the confines of our solar system; it is probably still sending its tiny messages, but by 2002 its signals were too weak to detect. Now it is well into galactic space and perhaps still speaking, but with no one able to listen.

Before the frantic race to build space-worthy vehicles began, engineers and scientists knew almost nothing about the environment of space. There were spectacular fiery failures of early space rockets before the rigours of the new medium began to be realized. I recall a vivid demonstration of the pitfalls that could entrap an innocent engineer. I was in a lab workshop discussing the design of a small gas chromatograph intended to be part of a Mars Lander, when a junior engineer asked, 'Can we use these fasteners for the oven assembly? It would save time and money compared with making them ourselves.'

'No, you can't,' was the immediate reply from his experienced colleague. 'Let me show you what will happen if you do.' He drew from his drawer a photograph of a cadmium-plated fastener that had been exposed to the hard vacuum of space; it showed how the metal had sublimed and re-condensed as a network of long, needle-like metal crystals. 'Those crystals,' he said, 'are like wires; they conduct electricity and will short-circuit any electronic equipment on which they form.'

Here was a whole new craft to learn, and we should look on the builders of those space vehicles and their instrument packages as the pioneers of this craft, just as were the Wright brothers and their followers the craftsmen of aeronautics.

In 1965, four years after I started with NASA, a husband-and-wife team of astronomers at the Pic de Midi Observatory in France, Pierre and Janine Connes, looked at the atmospheres of Mars and Venus through a telescope that had attached to it a remarkable new instrument called a Fourier Transform Spectrometer, developed by my friend Professor Peter Fellgett of Reading University in England. The spectrometer took all of the infrared light coming from the telescope and analysed the separate reflections of the different molecules in the two planets' atmospheres. When this data was interpreted at JPL by the astronomer Louis Kaplan, it showed that both of these planets had atmospheres made almost entirely of carbon dioxide, with only traces of oxygen and nitrogen present. I was enthralled because only a year before I had proposed that knowledge of the atmospheric composition of a planet would reveal whether or not it bore life. According to my hypothesis the atmospheric gases of Mars and Venus observed by the Connes implied that both planets were lifeless. This conclusion was welcomed by about half of the NASA scientists I worked with – the physicists and the planetary meteorologists – but was strenuously opposed by the biologists. Biology, after all, is about life, and if there was none on Mars they were out of a job. I saw it in an entirely different way: it gave us a new and wonderful insight into the nature of life on Earth. It has been the inspiration for my research ever since, as described in my first book *Gaia: A New Look at Life on Earth* (1979).

My experience at NASA in the 1960s highlights the interplay between science and engineering. The practice of science started long ago in some tribal community when someone thoughtful noticed, and remembered, a change in the natural environment that heralded something important for the food supply, safety or welfare of the tribe. The observation perhaps of high-level cirrus clouds after a long drought may have told of a coming monsoon and rain for the parched land. The faithful repetition yearly of this sequence of events would link the appearance of those clouds with relief from the long drought. This would then be tribal wisdom and could even be made long-lasting if the tribe had a written language. Probably such an observation would first have been made by an individual hunter returning at sunset or venturing out at dawn when the first appearance of the clouds would have been most visible. But engineering surely grew from the urgent needs of the hunter-gatherers: tools such as spears for hunting prey; cooking hearths and fireplaces; sharp implements for skinning and cutting meat. Then the tools for warfare and quarrels over territory must have exercised the brains of those primeval engineers and led to the design of better spears from wooden shafts and shaped flints. I suspect that the engineer's craft led also to better observation and strategic thinking.

Engineering and science are like a married couple that has never fallen out of love. Neither one could exist without the other: even a mathematician performing his intricate acts of intuition would be lost without a pen and paper; and who but an engineer could have made them? Where would the biologist be without a microscope? Galileo is remembered as a scientist, but he was first a superb engineer who designed and made a telescope with sufficient resolution to see the satellites of Jupiter.

We are wonderfully versatile animals. It may be convenient for bureaucrats, biographers and vice chancellors to categorize individuals as inventors, engineers or scientists, but in reality the terms are interchangeable. A good scientist is usually an inventor and an engineer as well, and most inventors are well versed in engineering and science. All of them have enough common sense to know that a girder should not be made from putty nor a hammer head from glass.

So down through history the wisdom of science accumulated. Lone observers noticed something different and wondered, then patiently waited and checked their observations to confirm that they always linked with the consequences of their prediction. Until about the middle of the twentieth century almost all science began this way; but then, especially in times of war, powerful leaders of governments and business imagined that science could be managed like an army. They were confident that the employment of a hundred scientists would achieve far more than one alone. Almost the reverse is true, in fact: even a million reasonably intelligent men or women gathered at the ultimate interdisciplinary conference would rarely if ever match an Einstein or a Darwin. Much worse, the funding of that million would leave nothing over to sustain a lone genius.

There were a few other lone scientists when I started to practise in 1964, but before long the numbers of them declined until now they are as rare as ectoplasm. Scientific journals refused to publish papers from a home address, and chemical and radioactive material suppliers would not sell to individuals. To overcome these barriers to starting work, I formed a commercial business called Brazzos Limited. The articles of incorporation were drawn up for a modest fee by a Salisbury solicitor: they allowed me to trade in any capacity, as my lawyer put it, ranging from a bank to a brothel. I think even he would have been surprised by some of the contracts that Brazzos Limited undertook, and still undertakes. My first source of income as a lone scientist was a consultancy with NASA at JPL, but my work with the space organization was soon channelled through a contract with Brazzos Limited. The next was to be a science adviser to Lord Rothschild, when he was science coordinator for the Shell Oil Company. He asked me and his other science advisers the question: what will the world be like for Shell in the year 2000? My answer was very different from that of the regular media prophets who, for the most part, were clinging to science fiction – such as forecasting colonies on the Moon or even Mars. Others, including Herman Kahn, predicted a comfortable world living a wealthy suburban existence, like that in Scarsdale near New York. And some offered gloomy images that echoed George Orwell's *1984*, of a dictatorship oppressing impoverished proles. It seemed to me that none of these projections would be true, although Herman Kahn's was closest: who would have guessed in the 1960s that India and China would become wealthy industrial powers? My guess for Shell was instead that by 2000, the Malthusian rooks would begin to flock in search of roosting places. The inevitable increase in population would be causing shortages of food and raw materials, and the environment would be increasingly polluted. So far you might think that I was thinking like a good Greenpeace follower, but I spoilt it by adding that in such a world, Shell could be doing very well. There are great opportunities in a degrading global environment for an enterprising green-washed energy industry, as German business has now discovered.

I soon found that the life of a lone scientist-inventor is like that of an artist or composer of music. Income is essential not merely for living but also to sustain a laboratory, and this can be quite expensive. It is therefore necessary to work for at least three providers simultaneously and do your real work in your spare time. To work for a single provider, no matter how generous, is merely to become again a bought man, an employee, and this is not independence. I stabilized with about five providers for whom I answered questions when asked. Even with five customers or providers, life is quite easy and leaves ample time to think about one's own work, in my case Gaia.

I even had spare time to investigate the CFCs, the detection of which at their parts per trillion scarcity was another triumph for the ECD. This small piece of lone science provided the data that led to the Montreal Protocol and the banning of the emission of these useful but dangerous substances.

My interest in CFCs started in 1970. I was puzzled by the dense haziness of the countryside air in the British Isles in the summertime, and noticed that the haze came with winds from the east or south, rarely with air masses from the north or west. I discussed this phenomenon with friends at the UK Met Office, then in Bracknell, west of London: they suggested that it must be an agricultural emission and were sure it was not air pollution coming from London. I also discussed it with Jim Lodge and other friends at the National Center for Atmospheric Research (NCAR) at Boulder in Colorado, USA. They expressed no opinion on the source of the haze, but gave me a sun photometer to use in England to make my measurements quantitative and comparable with similar measurements in other parts of the world. I started measurements as soon as I returned to my then home at Bowerchalke, Wiltshire, in the late 1960s, and they became a family routine that continued for several years, both in southern England and in the far west of Ireland. Soon it began to look more and more as if the haze was photochemical smog of the kind that plagues Los Angeles. We were amazed to find that the density of the summer haze in the rural village of Bowerchalke, nearly 100 miles west of London, was as great as that of Los Angeles.

To prove that what we were seeing was photochemical smog, I needed to show that the smoggy air contained a chemical that was unequivocally urban in origin and could not come from a natural source. It occurred to me that the chlorofluorocarbons (CFCs) fitted my need. They have no natural source and were used then as the propellant gas for spray cans. I could easily measure their abundance in the air outside my home lab in the Wiltshire countryside, even if it was as little as a few parts per trillion, with my newly invented ECD. I found these measurements solved my problem with the haze, proving that it must be photochemical smog, and a little later this idea was confirmed by a group from the UK's Atomic Energy Research Establishment at Harwell, who measured the ozone and other smog chemicals at my other haze observatory in the far western part of the Irish Republic. We soon realized that the presence of the CFCs in the air was a marker for an air mass coming from a densely populated region. In the 1970s they were the only measureable chemicals in the air that were unequivocally of human origin.

The air coming from the Atlantic Ocean contained far less CFC than that coming from Europe, but enough to make me curious about its fate and origin. I love ocean voyages on small ships and it occurred to me that if I travelled from England to Antarctica and back by ship I would be able to measure the amount and distribution of the CFCs in the Earth's atmosphere. At the time I was a visiting professor at the Department of Cybernetics of Reading University, an honorary task – no money changed hands in either direction – but it gave me academic credibility. My friend Fellgett, who was the real professor, and also a talented inventor of instruments, suggested that I apply to the Natural Environment Research Council (NERC) for a small grant to make the voyage on one of NERC's ships. The money from the grant was not to fund me or my research, it was for the benefit of Fellgett's department and to support a graduate student. This I did, but the grant proposal was vigorously, indeed rudely, rejected by the group of academics who advised NERC on whether or not to provide the proposer with funds. The letter of rejection read, 'Every schoolboy knows that the CFCs are among the most inert of chemicals, it would be difficult to measure their abundance in the air, or in sea water, as low as a part per million; the proposer claims to be able to measure their abundance at parts per trillion. The claim is bogus and the time of our committee should not be wasted by frivolous applications of this kind.' What a put-down! Fortunately, unlike a university scientist I did not need the money, and interestingly the covering letter from NERC hinted that the civil servants were not happy with the rejection and asked if I would let one of them visit me and see how a CFC measurement at fifty parts per trillion was done.

A few days later the NERC official came and watched as I measured the CFCs in the outside air. Then without further ado she said, 'We can't give you a grant to make the measurements – we have to accept the advice of your peers on our academic committee – but we can let you travel at no cost on the RV *Shackleton* on its way to Antarctica. You will have to supply your own equipment; all that we can do is to let you travel on the ship at no charge and provide a small amount of bench space.'

I was very happy with this decision. I did not want, or personally need, a grant, and I regarded Dr Howells, the NERC official, as a fairy godmother. She and her colleagues are often blamed for the errors of judgement made by groups of academic cronies who have the power to award or deny grant funds. To me, giving such power to the ignorant was a bad way to run science, but I do see that for routine applications, there may be some sense in awarding research grants on the basis of 'It's Buggins's turn next.' It is also scary to think that a lone outsider can in his spare time take some homemade equipment on a ship, make enough observations to show that there is a significant abundance of an industrial chemical in the air, and so cause the closing of a multibillion-dollar industry: no wonder lone scientists are not popular. The 'homemade' gas chromatograph which took me about three evenings to build is illustrated as Figure 4 and can be seen in the Science Museum in London. The measurements on the *Shackleton* went well and were published in the *Nature* paper

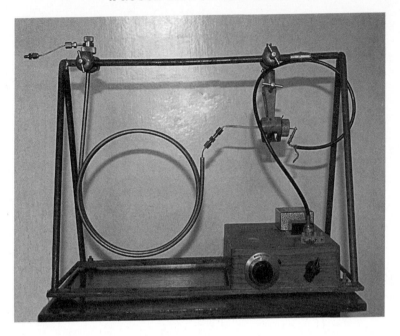

Figure 4. Homemade gas chromatograph.

'Halogenated methane in and over the Atlantic Ocean' in 1972. In *Homage to Gaia* there is a full description of the voyage, which I took only as far as Montevideo in Uruguay, and the way the measurements were diligently continued by the biologist Roger Wade, who was also doing research during the voyage, and the return journey measurements made by my graduate student, Bob Maggs.

Academics doing science at universities tend to think of themselves as professionals. Quite often they are as teachers, but less often do university scientists reach professional standards in their research. The technical sciences – physics, chemistry, biochemistry and so on – are taught with the graduation examination in mind. How often are undergraduates told when doing a practical test or measurement that the important thing to remember for the examination is the principle and the method – 'You will not have time to practise enough to become accurate.' In *Homage to Gaia* there is an account of an exchange between me, then an undergraduate, and my professor, A. R. Todd, in 1939 at Manchester University. Todd accused me of cheating in a practical chemistry examination. The grounds for his accusation were that students were never able to achieve an exactly correct answer in a practical analysis, and certainly never in a pair of analyses. Since I had twice given the right answer, I must have cheated. It took me ten minutes to convince him that I was professionally trained to do such analyses, and I demonstrated the next day how to do a more difficult analysis correctly. Todd was a decent man and he apologized for his mistake, but we both wondered about the quality of the students graduating from the chemistry department. Their conscious minds were trained but not their ability to perform well practically.

This distinction between academics and professionals almost led me to take legal action against the American National Academy of Sciences. In the first draft of their report on stratospheric ozone depletion by CFCs they stated that my measurements on the *Shackleton* were inaccurate. They made this foolish judgement mainly because my observations did not agree with their mathematical models of ozone depletion. They could have discredited in this way a university scientist, who is a paid employee, but to accuse a professional chemist of inaccuracy was, my lawyer told me, actionable. Fortunately, I was able to persuade the Academy to change the wording of the draft to something inoffensive. In the end the models were found to be incomplete and my measurements correct. It was the Earth, not me, that was confusing the measurements. My observations when the ship was near Europe were of air polluted by local sources of CFCs, and this made the north to south ratio of their abundance too large. The correct ratio was established by Jack Swinnerton, an officer of the US Navy, who made the journey from San Diego to the southern hemisphere on a research ship of the US Navy. His measurements were made using an ECD chromatograph that I built and supplied to him.

In the present controversy over climate change, too many scientists are either believers, who have faith in the theories of climate change, or deniers, who equally strongly do not. The same was true of the ozone depletion problem, and this was well described in the entertaining book by Lydia Dotto and Harold Schiff called *The Ozone War*.

This book coolly illustrated the extent to which many scientists were running over-rich on testosterone.

One of my aims in writing this book is to show that there still are lone scientist-inventors, although science is now biased against approval or support of them, so that it is difficult for them to succeed. In particular the recently devised processes of peer review and the funding of science by grant agencies are both prejudiced against outsiders and loners. There is no conspiracy against loners: peer reviewers and grant funding agencies try to be fair, but inevitably the membership of the committees that review applications for funds comes almost entirely from corporate science. The loners are unknown to them, or their ideas are so contrary to conventional wisdom that rightly or wrongly they reject them. The few lone scientists now in existence find it almost impossible to publish their work and ideas in approved scientific journals, regardless of its quality. Without peer-reviewed papers to judge an applicant, funding agencies cannot offer financial support. The lone scientist could be like his archetype, the artist, starving in a garret. That is bad enough, but rejection of the publication of science done by lone scientists too closely resembles the censorship of Galileo when loners are treated as heretics. But in truth, funding agencies are not really needed, and I'm glad that during the years I was a lone scientist I was able to avoid charitable help. In practice I found that, like an artist, I can live on 'pot boilers', the small creations that pay the living expenses. It is of course helpful to have a name before customers will buy these creations. I fear that as we move to a communal life in vast cities the automatic rejection of loners by the established teams will be seen as part of our evolutionary history and they may become extinct.

I have been a practising scientist for seventy years and a loner for fifty-six of them. I am not an impoverished amateur making gadgets in his garage, but professionally qualified in physics, chemistry and non-clinical medicine. In no way is this book meant as a polemic arguing for the replacement of teams by lone scientists. Nor is it meant as a denigration of teamwork. The truth is that there are places for large teams as well as for the lone creators. One has only to look at a cathedral to see this. St Paul's Cathedral in London makes a fine if obvious example; and its architect, Sir Christopher Wren, was also a

Fellow of the Royal Society of London, one of the world's first groups of scientists. The idea of the domed cathedral was Wren's, but it was built by a huge team of craftsmen. We need them both, and we need them now.

Yes, teamwork is needed, but those of you who have been members of 'think tanks' or have attended 'brain-storming sessions' will remember how rarely anything truly original or creative is achieved, other than agreement about the date of the next meeting. Great or dangerous ideas do come from that blue sky that seems to inspire leaders, but unfortunately only rarely. I can well remember being invited as an expert scientist to one of these brainstorms held by a UK government department. Also present was a humanist senior civil servant who asked, 'Why do we need these scientists when we can get it all on Google?' I sympathized with her because our scientific expertise had little to offer towards the purpose of the meeting, which was how London could be preserved as a peaceful city now that it was divided, apparently on ethnic grounds, into a set of potentially aggressive tribal groups. The members of our think tank were the best of scientists and administrators, but to me the lack of output suggested that teamwork was not a good route to discovery or original thoughts on that particular issue.

My message is a serious one, but I also want to offer an account of the fun and pleasure to be had doing science and inventing alone as it used to be done. We are forced by law to follow the often stifling rules of our, allegedly, egalitarian society; these have to be obeyed even though they are often little more than a triumph of bureaucracy over common sense. Pity the unfortunate scientist member of a team in a large laboratory. He needs some ultra-pure water for an experiment but to his surprise there is a health and safety warning on the bottle he receives. It tells him what to do if some of this ultra-pure water splashes in his eye, or spills on the lab bench. He must wash his eye and the bench thoroughly with clean water! Of course he ignores the warning but knows how unwise it would have been to make fun of the health and safety authority responsible for the warning. (This is a true account of the warning I received attached to a one litre bottle of deionized ultra-pure water a few years ago.)

The change we have experienced in moving from hunting and

gathering to our mechanized civilization today must have involved numerous encounters between groups within the tribes – groups we now call teams, committees, bureaucracies and so on – and individuals inspired by good and bad ideas. When working in Africa in the 1980s with my good friend Victor Pretorius, I heard a legend about an important tribe in Central Africa, the Masai. The legend claimed that a genius member of the tribe in the nineteenth century or earlier had the idea that cow's urine was the safest fluid for washing cooking utensils. Compared with the previous practice of using far from clean river water, it avoided the dangers of dysentery and probably saved many lives. This simple and effective public heath practice was cast out by medical missionaries who had quite different ideas, more religious than medical, about what was clean and what was dirty. Neither the original genius, nor the missionaries, knew anything about the epidemiology of water-borne disease. Whether or not there is any substance to this legend, it has stayed in my mind as a metaphor appropriate for many of our problems today. Inventions such as Newcomen's steam engine, Faraday's electrical machines, and the idea that fresh urine is a sterile fluid, all came long before their scientific understanding.

We might ask, now that the appalling threat of adverse climate change appears only to have been postponed, not ameliorated: should we encourage scientists expert in their different disciplines to work together in an alliance that made the best use of their talents? In other words, like the IPCC. If you think that this is self-evident, consider the statement of a prominent British politician some time before the last war. He said, 'To keep ahead of the enemy in scientific matters I don't need fifty PhDs in a chromium-plated laboratory; I need one long-haired genius.' It seems likely that the genius he sought was R. J. Mitchell, who designed the Spitfire, but by then he had died, although his ideas lived on and teams of engineers completed and replicated his design. But the politician was right, a genius was essential and no think tank or committee could ever have designed a Spitfire. In wartime it seems natural to couple the bright ideas of the few with the collective expertise of the many. Science and war are closely linked, and maybe this is why more often than not they are male preoccupations. It is time that we realized that most politicians and civil servants in peacetime have no capacity for problem-solving outside their immediate professional competence. They are too ready to assume that departments filled with scientific colleagues, paid experts and lobbyists will give them the right answers. Only rarely and then by accident does the answer came from an enlightened individual who happened to be a member of one of these groups.

My point here is that adverse climate change may be as serious and important as war, and we may need talented individuals as much as, or possibly more than, departments filled with experts. The civil service has a wonderful past record of dealing with human affairs in a discretionary and decent way but only rarely is it able to match the inventive capacity and leadership of truly able individuals.

War can be regarded as a way of naturally selecting the right ideas at a time when time is short. This is true enough for ideas in science or engineering, and the selection of the more effective usually happens, but on the social side a much more evolutionary way of selection appears to work best. Mistakes made in haste can have disastrous consequences. In my previous book (*The Vanishing Face of Gaia*) I described how the members of an unfocused human group can be so excited by an individual orator as to behave as if they were a single powerful individual. In a strangely similar way a collection of atoms or molecules in a laser can be excited from their resting state by an appropriate source of energy and then the energy they absorbed bursts forth in a single burst of great power. How like a lynch mob.

Rational scientists who are now sure that God either does not exist or is unimportant should wonder why they are so often believers of another kind, and hold informal and sometimes formal inquisitions to root out the heresy that denies their dogma. They should wonder why they ever allowed the convenient process of peer review to become established. It is a boon to editors, a great help for the many narrow specialties that make modern science so 'biodiverse'. But peer review can sometimes become a serious threat to the integrity of science itself and of its practitioners. Peer review has sometimes become the technique that cronies use to sustain their privileged positions. How similar it is to the Inquisition. Of course we do not burn scientist heretics at the stake, but they can easily be discredited and denied publication. This is no way to do science, and those who behave this way should be ashamed. Often, the believers are more probably correct, but we can't yet put a value to the probability. Above all we should never let science be the basis of a team sport between believers and deniers, because by joining a team you have given up your objectivity. Until the probability that there will be damaging or dangerous climate change reaches at least 90 per cent, let us keep cool.

So what is the lone scientist's life like? How does a lone scientist survive a lifetime of uncertainty and insecurity? For the first part of the writing of this book, the winter months of 2011–2013, I worked in a small room like the cell of a monk or a prisoner, the only furniture a desk and a chair. The computer gazed back at me and responded to the touch of my fingers; now that they are not so nimble, I rejoice in the excellence of Dragon Naturally Speaking, which magically transcribes my voice into text. The floor was carpeted in a uniform pastel shade close to the pale grey of the walls. Spare it might have been, but the room was kept comfortable by efficient air-conditioning that sustains at all seasons a temperature of 70°F. This was where I wrote. The room was on the first floor of what until 2013 was our winter home in Grover, a small community in the western suburbs of St Louis, Missouri. Unusually for suburb dwellers in the USA, Sandy and I had no car, and we walked daily the four miles to and from the shopping centre, bringing back the groceries in our rucksacks. For the few longer urban journeys we called taxis. There was no hardship to this way of life, and so low is the density of housing in this suburb that our daily walks to the shops had the quality of a country stroll, rich in wildlife and birdsong. If we were being green it was not by intention. We were motivated by a wish to stay healthy and enjoy an affordable life. We now live in a small cottage in Dorset on the coast of England in an area that has been declared a World Heritage site. We hope that its beauty and historical significance is enough to deter serious-minded politicians, who might otherwise see it as a site for more allegedly clean green renewable energy: clean it might be, but some of it is ugly and hopelessly impractical.

It is true that the working life I have just described is more that of a writer than an inventor or scientist. The change is recent, and up until about four years ago I worked in my lab, shown in Figure 3. My past customers are I think reluctant to entrust highly confidential or secret work to those past ninety years old.

In his book *Possible Worlds* (1927), J. B. S. Haldane has a chapter on amateur science. He stresses its utility and importance, especially its value in providing a way to do in a few hours and at little expense things that might take expensive months in the world of corporate science. But he warns of the dangerous tendency of amateurs to become believers, and to seek to prove their ideas instead of trying to falsify them. I fully recognize my own natural desire to try to prove myself right. On the other hand, being a loner gives a wonderful freedom to make mistakes and then learn from them; only very rarely do corporate scientists have this freedom, and the most dreadful blunders are too often made by large organizations whose members are unable to admit their errors. Worse is the human tendency of the team or tribe to cover up the blunder and protect the member who made it. Loners soon find that there is no one to cover up their mistakes; it takes a few years before we realize that this is the price of freedom, but for some of us, a price well worth paying.

I have often found personally that great adventures also come from what might be called the reduction to practice of sci-fi ideas. Audrey Smith and I had such an adventure in 1954 when we froze and successfully reanimated small animals (hamsters). The passage of time can temporarily be cancelled in two ways: firstly by suspended animation, survival in the frozen state, and secondly by travelling rapidly. If you could make a round-trip journey on a science fiction spaceship to the locality of our closest star, in the Centauri group – 5 light years distant from the Earth – you would find on your return that everyone had aged ten years, although, thanks to relativistic time dilation, everyone on board the spaceship would have aged by a lesser amount – how much less depends on how fast the spaceship moves. These two procedures are, for now, both science fictional, at least for people. Hamsters are, I think, the only mammals to have been frozen and reanimated. Humans probably are too large to survive the stresses of freezing and thawing.

Another adventure happened when I was inspired by the thoughts of that great nineteenth-century physicist Clerk Maxwell. He imagined a minute demon, who at a tiny door between two containers could sense the arrival of a fast-moving molecule and bat it back, but let a slow-moving one through. In time one container would fill with fast-moving (therefore hot) molecules and the other with slow-moving (cold) molecules. A heat engine could be connected to the two containers, and energy could be had for free. Apparently, the demon had committed the unpardonable and heinous offence of breaking the second law of thermodynamics, the most fundamental law of the universe. In real life the demon could not have separated hot from cold molecules without using as much energy or more than the heat engine could deliver. Information about the nature and properties of an atom cannot be had for free, and like all of us the demon would have to work for it.

Despite being warned that I would be wasting my time and NASA's money, I had a great deal of fun at the JPL in the 1960s making a relative of Maxwell's demon, the segregating demon, one that could be the gatekeeper between two containers each full of a gas mixture, and completely separate the two gases so that each of the two containers were now filled with pure but different gases. This time it worked, and its practical form was used on the *Viking* landers that went to Mars in 1975. The segregating demon was not cosmically criminal; it had merely found, as does a good accountant, a legal way partially to avoid (not evade) the tax imposed by the second law. This demon worked as an essential component of the gas chromatograph mass spectrometer instrument that analysed the Martian regolith and atmosphere. Three plaques from NASA acknowledging this invention and its use on Mars are now on the wall of my workroom by the sea, but for me the best reward is to look at the sky above the sea and sometimes see the red planet, and know that two of my inventions landed safely on its regolith and worked as they were expected to.

Lone scientists are sometimes able to provide an independent viewpoint for politicians and senior administrators. One evening in the middle of the 1970s I received a phone call from Dr Lester Machta, a senior scientist at the US National Oceanic and Atmospheric Administration (NOAA) and head of their Air Resources Laboratory near Washington, DC. Without preamble, Lester said, 'Jim, we have a truly worrying problem that I would like to share with you in the hope that you can throw some light to help us find a solution.' He then explained how NOAA had set up a network of instruments to measure the intensity of solar ultraviolet that reached the ground at airports around the USA, from Alaska to Florida. These stations had been monitoring solar UV for the previous ten years and to everyone's consternation the instruments recorded a steady decline in UV intensity, not the increase that had been predicted by the computer models of ozone depletion in the stratosphere. This was the holy grail sought by environmental investigators of those times. And as it is now with 'global warming', there were believers, deniers and sceptics. The believers in the 1970s were certain that the destruction of ozone in the stratosphere would cause an increase in solar UV at the Earth's surface, and they were sure that since CFCs were steadily increasing in abundance, the UV observed at the surface must also be increasing. They were so sure about it that if these instruments said that the UV was decreasing not increasing, they concluded there must be something wrong with them. Since the leaders among the believers included powerful politicians and administrators, Lester was worried and had called me because I was a scientific instrument designer and might know why the NOAA instruments were failing to give the politically correct answer.

My reply was, 'Lester, I would not like to be in a plane with you as the pilot on a dark, foggy night. Why do you prefer to assume that the instruments are failing, and all of them progressively? Perhaps the UV reaching the ground really is declining.'

When the UV monitoring procedures were examined by professional investigators, they confirmed that the instruments were correct and that UV was indeed declining. The true reason for its decline was the progressive increase in the pollution of the northern hemisphere air with aerosols, and tropospheric ozone, both of which absorb or scatter UV from the sun and so reduce the amount of UV that reaches the ground. Models can be misleading, especially when made and interpreted by believers. A similar confusion is taking place now as believers in global warming are discomfitted by the failure of the real climate to agree with model projections. As mentioned in Chapter 4, the contemporary arguments about climate may also be in part attributed to a failure to include the intricate effects of small particles and droplets floating in the air – what scientists call atmospheric aerosols.

This autobiographical chapter is included because it helps explain the unusual viewpoints from which I have had the chance to observe the changing scene. The scientists whose lives have most inspired me are Michael Faraday, Henry Cavendish, Alan Turing and John von Neumann – all four of them talented inventors as well as scientists. I have also been inspired by the writings of my friends, who have sensed the revolutionary change in evolution that is happening to us and which is wholly different from the myth of progress.

3

How Invention Accelerated Evolution

This chapter is about the part invention plays in the evolution of the Earth and all of its inhabitants, including us. For the Earth system to act as a living planet capable of self-regulation it has to be populated with a sufficient number of living organisms. A group of astronauts on the Moon does not make the Moon a live planet: it is a large, dead piece of rock and the temporary biomass of the visiting astronauts, despite their huge content of bacteria, plays no part in regulating the physics and chemistry of the Moon. As we shall see, so long as there are sufficient organisms living and participating with the surface physics and chemistry of a planet, the biological form of these living organisms may not be important. This is strikingly illustrated by the difference between the anoxic (that is oxygen-free) world of the Earth's earliest biosphere, the Archaean, populated by micro-organisms alone, and the oxygen-rich atmosphere we now enjoy, populated in addition with plants and animals.

The Earth was formed about 4.5 billion years ago and endured a period called the Hadean, of great violence, culminating in what is thought to have been the impact of a planetary body as large as Mars. The energy released by this collision is thought to have melted the combined new planet and somehow from the molten mass the Earth and Moon were formed. This period continued until the Earth was cool and stable enough for life to emerge, commonly assumed to have been at about 3 billion years ago. There is very little hard evidence of those ancient times, and we are forced to speculate that there was little or no oxygen in the air and the Earth was probably rather cool. Our reason for thinking this is that stars like the Sun grow hotter as they age, so the heat received by the Earth then was probably 25 to

30 per cent less than now. To sustain viability on this cooler planet our distant bacterial ancestors were likely to have been naturally selected after they first appeared to favour those that warmed their environment. Among their inventions were biochemical methods for turning the dead bodies of early bacterial life and other organic detritus into the greenhouse gases methane and carbon dioxide, which helped create a more benign world for the selected bacteria. To those first organisms this was an intricate task, and it is likely to have involved advanced technology that included quantum as well as traditional physics and chemistry. Thus the more energetic photons of sunlight, those associated with what we see as the blue and violet colours of the rainbow, are capable of severing chemical bonds. The less energetic photons in sunlight, those of the red and infrared, supply mere warmth and are unlikely to provide conditions in which life could easily appear or sustain itself.

I like to speculate that Archaean life had a form of Internet on which the recipes for their clever chemical tricks could be passed on as downloadable information – the earliest apps. The archaic Internet would have used the air as its medium: tiny programmes, plasmids, written in DNA or RNA, would have been carried by the wind as aerosol particles around the Earth in one of the two regions called hemispheres that exists between the Equator and the Poles. Travel around a hemisphere could be in as short a time as eight days. Travel between the northern and southern hemispheres is much slower and may have taken a year or so. Like us, they ended up with quite a complex planetary ecosystem. The need for planetary self-regulation of climate and chemistry probably was as great then as it is now. In the Archaean period it seems probable that a planet-wide self-organizing life form regulated the planet's climate and chemistry. For example, the recycling of chemicals was essential and it is not difficult to think of scenarios in which the carbon dioxide in the air, the gaseous blanket that kept those prokaryotes warm, would have been drawn down by reactions in the sea, or at the surface of the rocks, and so threatened the Earth with freezing. Once freezing started, it would unleash a runaway positive feedback leading to further freezing. The Russian meteorologist Mikhail Budyko rightly realized in the 1970s that land permanently covered by snow would reflects so much sunlight back to space that temperatures would fall further and the area of snow cover would expand until possibly the whole Earth would be snow- and ice-covered. A snowball Earth could have been stable for a million years or more with the cooler Archaean sun until, as geologist James Walker suggested, carbon dioxide from volcanoes accumulated in the air sufficiently to make a greenhouse thick enough to melt the ice. The invention by prokaryotes of a way to synthesize methane, a green- house gas more than twenty times as effective as carbon dioxide, may have taken place in one of Darwin's 'warm little ponds' – the habitats he envisaged that brought together primeval chemicals and micro-organisms. I suspect that we would not be discussing it now if this little app for making methane had not been written in DNA, and then posted on the wind and shared with everyone. These small inven- tions emerged to meet both large and small problems of Archaean life and enabled the first working form of a living and self-regulating planet. Methane in the air now does not last long because it reacts in sunlight with oxygen so that about 67 per cent of it is destroyed in ten years, but in the Archaean there was hardly any oxygen and only a small production of methane would have been sufficient to create and sustain an effective greenhouse. Ten to 20 parts per million of meth- ane in the Archaean could be expected to warm as much as 1,000 parts per million of carbon dioxide. Methane then was a near perfect envir- onmental invention: not only was it more than twenty times as effective a greenhouse gas as carbon dioxide, but its decomposition products in the stratosphere would have formed an effective UV shield – as powerful (or more so) than the ozone layer that came when oxygen dominated the air. Confirmation of this idea of the methane equivalent of the stratospheric ozone layer comes from observations of the atmosphere of the satellite of Saturn, Titan. It is rich in methane and the decomposition products form a brown haze that is strongly UV absorbing.

We do not know enough about the early history of the Earth to be sure about who were its first residents. It seems likely that they will have evolved rapidly and that among their progeny there will have been an abundant group of organisms capable of forming a system that sustained a habitable planet. The Earth's first housekeepers in this sense were the prokaryotes, bacteria that emerged at some time before 3 billion years ago and appear to have sustained a planetary environment chemically dominated by the simple gas, methane. It was a world where oxygen was a minor constituent and it lasted until about 2.2 billion years ago, when oxygen became abundant enough to drive the anaerobic organisms underground. Without oxygen to breathe we die in minutes, whereas to the world of anaerobes it is a deadly poison: but surprisingly, a layer of soil less than a millimetre deep is enough to shield anaerobes from poisoning by oxidation. After the Archaean, oxygen became the chemically dominant gas of the atmosphere, and has remained so until now. In the next age, the Proterozoic, there was enough free oxygen in the environment to make the surface of the Earth oxidized, but nowhere near enough for large animals to flourish. It lasted until about 700 million years ago when large plants and animals appeared: as they grew in size oxygen increased until it reached the current 21 per cent abundance. We must never forget that we and the animals of our world are dependent on oxygen at near its present abundance. The earlier worlds would have been so inhospitable that we would have fallen unconscious after breathing their air for as little as twenty seconds. The air then was a mixture of nitrogen, carbon dioxide and methane: not poisonous, but so lacking in the oxygen we need that we would have been dead within a few minutes. We should also keep in mind that our oxygen-rich atmosphere suits us but may not suit our successors.

What I have just said about the history of life on the early Earth is not much more than educated speculation: the reality could have been quite different. It may, alternatively, have been that the early Earth was almost entirely covered by sea-water rich in the salts of elements like iron, nickel, manganese and others. Seen from space such an Earth would have had a darker sea than now, and the planet would have absorbed nearly all of the Sun's heat radiation in the upper layer of the ocean. If these were the conditions 4 billion years ago, then cooling rather than heating might have been needed to bring the surface environment to one near to optimal for life. The most up-to-date report on the Earth's history is in a concise and pleasing book, *Revolutions That Made the Earth* (2011), by my colleagues Tim Lenton and Andrew Watson.

It is awesome to think that those messages, those apps the

prokaryotes exchanged so long ago, are still with us in our genes and in those of all forms of life on Earth. The information in the fragments of DNA and RNA, when decoded, might tell us more about our origins than can be found in the few Archaean rocks that still remain. But it is no easy matter to read the message, the story of life's origin about 3 billion years ago, from the account laid down in our genes.

I started this chapter on inventions with an account of the early Earth because of the insight it gives on the origin of other living inventions, such as the eukaryotic cell that hugely changed the course of evolution, and of course into the artefacts and the inventions that we have made.

We are usually taught that human history moves back seamlessly for thousands of years, at least as far as something memorable like Stonehenge, or the Pyramids of Egypt. But this book argues that there was a huge discontinuity that began at the start of the eighteenth century that made the evolution of our artefacts accelerate to a pace far beyond the capabilities of natural selection. It has become so rapid that it is more like the physical inflation that occurred early in the evolution of the universe than Darwinian evolution. The reason why we did not notice the start of this new form of evolution was, I suspect, due to the fact that when caught in the grip of purely physical events, you are often unaware that anything is happening. On the few occasions when I was a passenger on Concorde I was wholly unaware that I was travelling faster than a bullet from a high-powered rifle. Of course all of us are travelling with the Earth in its orbits of the Sun at 16 miles per second, or 57,000 miles an hour, but there is no sense of motion. Physicists say that we might even be unaware of travelling at a pace that approached the velocity of light when crossing the event horizon of a large black hole. Provided that the initial conditions of our journey were suitable, we would be unaware by sensation that we had moved from a place outside, where escape was still possible, to one inside of utter disintegration. The man who jumped from the twentieth floor of a skyscraper said, as he passed the third floor, 'It's OK so far.' A few of these jumpers even survived relatively undamaged when they encountered not the hard pavement but the thin roof of a car or van and decelerated slowly enough to avoid serious injury. Are we now passing our metaphorical third floor? Take comfort also in the thought that the universe survived its inflation; perhaps we will too.

It is fashionable now to regard us as guilty and attribute to us the harm that came from industry. But what if instead it happened quite by chance, and although we humans and the whole biota have participated, it started spontaneously like a thunderstorm and no one is morally responsible? Neither is it necessary to invoke an act of God or some other external transcendental force for its origin. I think it possible that we humans in our everyday business accidentally and unintentionally set the scene for its emergence. Always keep in mind when seeking someone to blame that none of this 'carbon problem' would have happened if early micro-organisms had failed to split water molecules and so release the oxygen. Without abundant oxygen neither animals nor fires can exist.

Was it the discovery that fire could be useful that set us on the path to the world of today? Certainly not: invertebrates beat us to it by at least millions of years. The first to employ combustion in a purposeful way was the remarkable bombardier beetle, which has an internal combustion chamber like, only much smaller than, that of a rocket. The insect pumps fuel and oxidant into the chamber and causes an almost explosive hypergolic combustion that makes a bang and a jet of hot steam laden with irritants. This is how it kills or scares away predators. It still fires its shots in defence but has never seized the chance to evolve as an invertebrate space traveller.

The first group of humans to use fire purposely for cooking, and to deter predators, had an advantage in the contest for selection. Cooking reduced their burden of parasites, rendered harmless both vegetable and bacterial toxins, and not least promoted itself through its novel flavours and tastes. According to Richard Wrangham in his book *Catching Fire* (2009), cooking allowed our brain size to increase, which presumably added further to our evolutionary advantages.

The increase in population consequent upon this use of fire may have driven the shift from hunting and gathering to farming, which in its turn brought huge advantages and a further growth in the number of humans. The needs of farmers soon led to the formation of villages and then to cities, and these in turn gave birth to civilizations. All of these collections of people continued to burn fuel for cooking and to keep warm in winter, but not at a rate sufficient to change the surface chemistry of the Earth.

We are fairly confident that what we call 'civilizations' go back to a time 7,000–12,000 years ago; but fire-using humans had been around for at least 1 million years. Perhaps they used fire in that legendary civilization Atlantis, and perhaps it was not the only civilization to be inundated as the ice melted and the oceans rose. The legend of Gilgamesh and the story of Noah's flood may be related to the melting of ice at the end of the most recent Ice Age and the inundation of civilizations as the oceans rose close to 100 metres. Our present cities and their artefacts can be expected to suffer a similar but lesser inundation as global warming melts the remaining ice of the two polar regions. If it all melts, the sea level could rise about 30 metres, more than enough to bury today's civilization, except in those towns above the high-water level. Mexico City, Denver, Geneva, Rome, New Delhi, Islamabad, Johannesburg and Nagoya come to mind, and the world's highest city Wenquan, is in China: but would these survivors, when disconnected, retain the industrial civilization we now experience?

I find it extraordinary that despite being almost our equals in science, mathematics, literature, art and philosophy the Chinese, Greek and Roman civilizations never climbed the crucial first steps that led to the modern industrial world. Why, for example, did the Romans have such impressive civil engineering, with their straight and perfectly cambered roads, sturdy bridges, aqueducts and splendid architecture, but no railways, cars or air travel? Why did none of them invent the steam engine?

Perhaps it was as Newton said, when disingenuously denying his genius: 'If I have seen further it is by standing on the shoulders of giants.' In a way, we did the same: maybe our extraordinary achievements came about because we built on the infrastructure laid down by those older civilizations. There must be some truth in this, but I think that something much more powerful happened. About three centuries ago we unintentionally set in motion an accelerated evolution that now grips us and our world in a series of unstoppable events. We are like the sorcerer's apprentice, trapped in the consequences of our meddling. Make no mistake, the accelerated evolution of our artefacts is no mere hyperbole: it is a solid fact of observation. That neat new telephone in your pocket can communicate even visually with others almost anywhere on Earth. It allows you to send as a live movie the scene before your eyes, to be immediately seen by someone on the other side of the world as if the distant friend was your neighbour next door. An intelligent Roman 2,000 years ago would have regarded this as magic, and that is what it is to most people today. (As Arthur C. Clarke put it, 'Any sufficiently advanced technology is indistinguishable from magic.') Almost certainly, neither you nor anyone you know has more than a vague idea how the phone works, or could make one.

We humans, and all else that is alive on the Earth today, have taken more than 3 billion years to evolve by the slow process of Darwinian natural selection. So what caused the rapid recent rate of inflation in the evolution of our artefacts? Something that happens far faster than the usual rate of evolution by natural selection. As I mentioned in the first chapter, the evolution of the universe itself did not proceed as a slow, smooth and constant process; instead, soon after the Big Bang there was a period of incomprehensibly rapid physical evolution called 'inflation'. It happened so rapidly that the parts of the inflating universe may have moved apart faster than the speed of light. The cause is still the subject of speculation, but there is ample evidence that it happened. You may have seen photographs showing a map of the universe shortly after the Big Bang had occurred. It is usually an oval-shaped map somewhat like those maps of the Earth shown to illustrate global warming.

Most cosmologists think that the universe started with a Big Bang. All of the mass of the universe was compressed into a tiny space from which it expanded at a furious pace. The intense compression of a whole universe into something as small as an atom made it unimaginably hot, and from this state it expanded at a pace faster than the speed of light until it cooled to a temperature low enough for particles called quarks to condense; eventually a few minutes after the bang the temperature was about a billion degrees and the density about that of air at sea level. After this it continued to cool and after nearly 400,000 years became mostly hydrogen, helium and lithium. From this hot gas eventually stars formed, and in 13.8 billion years we have the universe we live in now. The intense radiant heat of this hot material universe also expanded and cooled until now it is only 2.7 degrees hotter than absolute zero (less than 273 °C below the freezing point of water). This is relict heat radiation, known as the cosmic microwave background radiation. It was discovered accidentally when two physicists, Arno Penzias and Robert Woodrow Wilson, were testing a new radio telescope. You can see the presence of this radiation yourself as a scattering of luminous spots on the screen of your television early in the morning when you are far from a city and all TV channels have shut down or are too distant to be seen or heard. We now know that some of the apparently random flashes on the screen are records of the relict radiation of the Big Bang. Alan Guth first conceived the idea of cosmic inflation as a way to explain the facts of the universe in the late 1970s. The form, intensity and existence of the microwave background noise confirmed Guth's inflation theory so well that most physicists now take it as fact.

Cosmic inflation stopped and left us with the universe we now inhabit. The accelerated evolution of our inventions, like all exponential growth, must inevitably level off or stop, and the signs are there that climate change, population growth and economic growth – which are all linked to the inflation of inventions – are, all of them, showing signs of levelling off. I am not suggesting that the accelerated evolution of our inventions that we experience now and Guth's cosmic inflation are directly connected, but these two hypothetical phenomena share a common connection through the fact that they both concern the flux of energy. It was a sufficient flux of energy, which came and still comes in the form of sunlight, that enabled the appearance of life on the Earth; and the runaway invention of the Anthropocene happened when the energy from a coal fire was great enough to sustain it.

Whenever the flow of energy passes well-defined critical values, something unexpected happens. A simple, everyday example is what happens when you turn on a kitchen tap. If you do it slowly, at first the flow is a steady, smooth stream that runs quietly into the sink. If you now slowly open the tap the flow increases and then quite suddenly the smooth stream breaks into waves and ripples that move down the column of water. Give the tap another turn and the flow becomes a noisy torrent. The emergence of unsteady flow in the water stream is what we call turbulence, and the conditions for its appearance are precisely set by the properties of water, by the shape and diameter of the pipe from which it flows, and by the rate of flow. It happens in all fluids and most of us have encountered it as the annoying turbulence we experience when flying in a plane. The Victorian engineer-scientist Osborne Reynolds first reported it, and his name is attached to the number that marks the onset of turbulent flow. (Scientists sometimes choose to take an old word like 'fluid', which we all took as a synonym for liquid and broaden it to include gases like air or carbon dioxide.)

Eddies and vortices that appear in a flowing stream of gas or liquid are examples of this phenomenon of turbulence. In some ways the vortices resemble forms of life: they reduce their internal entropy but excrete entropy to the environment. They are born and die, and even give birth to smaller eddies – as the famous meteorologist Lewis Fry Richardson once said poetically, 'Big whirls have little whirls that feed on their velocity, and little whirls have lesser whirls, and so on to viscosity.' Truly large whirls – hurricanes – are so lively that we give them individual human names. Could it be that the even more lively appearance of life itself on the Earth, or on any other planet, is linked in a similar way to energy flow in a natural system? And does life require a similar set of critical physical parameters for its onset?

You may think me tedious to keep writing about the flow or flux of energy in connection with life and inventions. If you do, I apologize for failing to convey the excitement I feel about the fact that the turbulence of a flowing stream and life itself share membership of the same family of physical phenomena. Life and inventions are far too important to be taken for granted and for these reasons:

1. Life is a dynamic process that when alive is in constant motion, and never static.
2. Carbon life can only exist between the freezing point of the ocean and 50°C, and to keep within this range a planet needs a continuous flux of energy from its star, whether it has life or not.
3. The presence of life on a planet affects the flow of energy needed to sustain it, the Gaia effect.

We really do need to view the broad range of factors involved in the energy balance of the Earth and other planets. It is not enough to limit it to any one or two of the academic disciplines. The broader view affects our thoughts on such diverse topics as climate change, exobiology, the politics of sustaining the environment, and so on.

For example, exobiologists seeking the presence of our form of life on distant planetary systems use the term 'zone of habitability' to define the region around a distant star in which life might be found. Unless the concept of Gaia is accepted, the idea of a zone of habitability is limited to a simple physical range of orbits within which the star's heat radiation would keep an inert planet within the range of temperatures consistent with life. This narrowly limited range of orbital radii is sometimes called the 'Goldilocks' zone, by analogy with the nursery tale about Goldilocks and the three bears.

We do not know accurately the smallest flux of energy to the Earth that could have started life but, as mentioned earlier, we do know that when life started on our planet the Sun was supplying about 1 kilowatt of energy to each square metre of the Earth's surface. We are also fairly sure that life can only form in a liquid medium, and since water is the dominant Earth liquid, life could only form above its freezing point of 0°C and below its boiling point at whatever was the atmospheric pressure. In fact, the upper limit is likely to be 50°C. You can check this yourself by placing hot objects on your skin: objects at or above 50°C will burn you, whereas below this temperature they will feel hot but bearable. This is true for nearly all life, and is why milk and other foods are pasteurized at 60°C to kill pathogenic micro-organisms. 60°C is a safe ten degrees above the normal lethal temperature for micro-organisms, but not so hot as to spoil the taste.

The first life on Earth may have originated from the assortment of chemicals left around soon after our planet formed, plus others made by inorganic chemistry and photochemistry. The temperature must have been somewhere between freezing and 50°C. Many of the random chemicals from space would have reacted to form more complex substances, but if you intend to build life from the spare parts of a cosmic junkyard – to allow evolution – it is not enough merely to join chemicals together. You would also need a mechanism to break them down and start again or make something quite different. The early Earth had several cutting tools available for this job, provided by natural radioactivity. Our planet is a by-product of a nearby star-sized nuclear explosion. At the time life began, it was still hot enough in a radioactive way for anyone so inclined to make nuclear weapons from naturally occurring uranium. In addition to this natural supply of high-energy radiation, sunlight at the blue to violet and ultraviolet end of its spectrum provides radiation powerful enough to break chemical bonds. It's important to appreciate that those radiations, nuclear and solar ultraviolet, were crucial for life's formation and continuation. Chemical bonds can be severed also by heat, but mere heat is less selective about the bonds broken than is radiation, which is why heating to 60°C is usually fatal to all except for a few rare species. Chemical bonds can also be broken by reaction with highly reactive substances called free radicals, and these can be made at the temperatures associated with life. One of these free radicals, hydroxyl, the -OH radical, acts like radiation in bond-breaking. It is often the first product of radiation, and important in its harmfulness. But the production of -OH by radiation usually requires the presence of oxygen, and the air of the early Earth is unlikely to have had any significant abundance of oxygen.

I recall discussing in the 1960s with the astronomer Carl Sagan the possibility that life might originate on a planet orbiting a small red-hot star, cooler than our yellow-white Sun. Carl thought that it could start on such a planet if it was near enough to its star to have a similar climate to the Earth and was made from the same components. I thought that the chances of life beginning near a cooler star were much less because the red star would not emit high quantum-energy radiation; mere warmth alone is not enough for photosynthesis. The bright light of the blue and ultraviolet from the Sun is needed to kick-start life, and for undoing the chemical bonds made in error.

Life, the greatest invention of all, came about by chance, as Jacques Monod explained in *Chance and Necessity* (1972). Monod's words reinforced for me my trust in science and rendered unnecessary any need to invoke personalities like God, or wandering astronauts from elsewhere in the universe. Understanding the origin of life gives insight into the origin of other living inventions, such as the eukaryotic cell that hugely changed the course of evolution – and, of course, the artefacts and the inventions that we have made. Just as the appearance of life on Earth profoundly changed our planet's evolution and future, so the surfeit of invention is now changing our and our planet's evolution in ways far stranger than can be imagined.

The Anthropocene could not have happened without humans, so what is it that distinguishes us from all other forms of life on Earth? We can stand upright leaving our arms free; our hands have reverted to the original reptilian design and have five flexible digits now adapted as and for tools. We have acquired the ability to walk or run great distances. We can even outrun a horse, because it is not speed that makes us strong, but endurance and the will to keep going until after the horse tires; our intellect then out-thinks the horse, anticipates its escape plan and catches it. We can gather in groups and share ideas, but we also know that invertebrates apparently as simple as the bee can discuss the best site for foraging for honey, and more remarkably, discuss in a swarm the best site for the new nest. We can transcribe our words and keep the story for a longer time than any individual can live. But what makes us special is, perhaps, something more than the combination of all of these beneficial evolved traits; nor is it one single organ like the brain.

The extra advantage that makes humans different and astonishingly powerful is the way our minds and bodies interact: the close coupling of our brain to our hands, and indeed any other mobile part such as our voices. This facile link with the brain makes available for us a range of peripherals that we use for communicating. We can also invent tools and weapons, and then practise until we can use them as fluently as a violin player in the midst of a passionate sonata. Without tools, hands to use them and a brain to direct and compose, we would be trapped like the whales in that vast and re-entrant ocean that is their home and prison. We have evolved to become inventors.

Our ancestors 10,000 years ago, as Arthur Clarke pointed out, would have regarded our daily activities now as magic; yet we hardly differ from them in any important way. We and they have evolved by the rules of Darwinian natural selection: the organism that leaves the most progeny is the one selected. This is an extraordinarily slow process; just how slow is shown by the time it took the first green living things able to turn sunlight into food and oxygen, and the cyanobacteria of the ocean, to evolve and become plants and trees. Like our own biological evolution from the first cells that appeared on Earth, this took about 3 billion years, a sizeable fraction of the age of the universe, which is now thought to be 13.8 billion years.

Until about the eighteenth century, we and all life on Earth were still evolving at the snail's pace of natural selection. Then, quietly and with no fanfares, we started another form of evolution, one that increased in pace exponentially. We are still naturally selected human animals, but we are now doing things that would have been effectively impossible before this new evolutionary inflation started. Things such as using the heat of a coal fire to travel by rail, or make electricity. What would an intelligent Roman have made of your Blu-ray DVD player as it reads the pits burnt into the surface of a plastic disc and then displays in full colour the lively action of a drama on a life-sized high-definition screen?

In a way similar to that which led our forebears to believe that all the marvels of the Earth and life upon it came from the hand of God, we think that these recent marvels are all the work of the brains of great engineers and scientists. We have been led to believe that the modern world is the norm of our scientific age, and we take it for granted, just as our forebears did the work of God. There is an ever-growing number of what we regard as common objects that our hypothetical intelligent Roman would have found ineffable: things such as cooking in a microwave oven, using the Internet and travelling through space as far as the Moon and back. Then there are more mundane things – like detergents, medicines that actually cure disease, and rope made from polypropylene, a plastic so light that it floats on water yet is comparable in strength to steel.

But the invention of these artefacts takes place without more than a tiny proportion of humanity having the slightest idea of how they were made or how they work. This is even true of the inventors; only rarely is an invention rationally understood by the inventor at the time of invention. I know, for I am as much an inventor as a scientist, that it can sometimes take years to partially understand or explain the invention that came into my mind almost instantly as a gift from intuition. Scientists, unless the invention is a part of their speciality, are as ignorant of the working of our artefacts as the rest of us. How many practising scientists can explain in full how their cars or their laptops work?

I used to worry when I visited a university, government agency or large commercial firm and saw labs filled with intricate and expensive instruments. It all looked most impressive, but I was concerned because I knew that those who used the instruments only rarely knew how they worked, or could have made them. Worse, the scientists who made the measurements seldom knew if their instrument was in error. It is similar to your own relationship with your car. You know how to drive, and if the steering mechanism errs you sense the error and compensate. But if the complicated instruments in the lab are in error it may be a long time before the scientist becomes aware of the error and the fault is recognized. I am quite sure about this because I had helped a large instrument company, Hewlett Packard, design a laboratory gas chromatograph and then seen it in use in the laboratories of universities, government departments and commercial firms. Far too often errors of measurement as large as hundreds of per cent went unnoticed and became part of the conventional wisdom of science through publication in peer-reviewed journals.

My target for denigration is not science or scientists; after all I am one myself. My target is our obsession with the power of rational thinking that goes in parallel with our distrust of intuition. Intuition and reason are both part of our evolutionary past and I suspect are equally necessary for survival. As both an inventor and a scientist, I learnt early the first rule of a scientist's life is 'publish or perish'. This forces those who do science to describe their researches rationally. Peer reviewers tend to reject papers that speculate about the discovery that the authors have made, even when it is important. Better to appear ignorant than offer what is seen as an irrational explanation.

Inventors do not need to publish in peer-reviewed journals. They invent their widget and then describe what it does to a patent lawyer, who then files a patent application. This is a description of their invention and what it does, but not an explanation of its principles or how it works. The patent attorney expresses the invention in recondite but comprehensible language. The patent is usually headed by the phrase 'Improvements in and relating to . . .' Patent law is mainly concerned with establishing priority. It cares not at all about a rational explanation of how a widget works and is content with phrases like 'Well known to one skilled in the art . . .'

The point of this digression about science and invention is to acknowledge that while these two skills occupy two different cultural worlds, they are both held comfortably in our brains and both have been naturally selected by chance and necessity. Most scientists are inventors and many inventors are scientists; what difference there is between them lies in the way they think and communicate their ideas, and the outcome they seek. The inventor usually seeks to answer a need and receive a cash return; the scientist seeks recognition, status and a salary. This is well illustrated by my own professional life, described in Chapter 2. The exponential growth of artefacts in the last 300 years was mainly through inventions, and these were not inspired by science but by necessity. The part played by science was to offer explanations, only rarely to invent. If you really need someone to blame for global warming and all that, blame the inventors, not the scientists.

I like to think that while inventing I am in touch with that inner layer of the mind where information is processed without awareness. As a scientist I have to explain each step of a process openly, and often I do this on a piece of paper, with pencilled sums or simple diagrams. In real life these two processes tend to merge, and scientific intuitions come from the inner layer without my being aware until the completed thought emerges. It is also true that some inventions can be devised rationally.

As I mentioned earlier, no exponentially growing process goes on forever. It is probable that accelerated evolution will cease or change its pace some time in the present century. I suspect that we can already feel the brakes applied to the rate of change of climate, population and money supply. Just as well, for otherwise we and Gaia might be trapped in the truly ineffable world of a singularity, as the futurists Vernor Vinge and Ray Kurzweil have warned. The most probable cause of a deceleration is the rising cost of energy.

We joined this accelerating conveyor belt at a time when a few inventors realized that the heat energy of a coal fire could be used to boil water and the steam used to drive engines. At first we humans did not evolve more rapidly than before. But because we are tightly coupled to the evolution of our artefacts, our biological evolution and that of the whole Earth system is now dragged along with it – more slowly than the evolution of artefacts, but still for us at a breakneck pace. As invention begins to invade the biological sphere, with genetic engineering and even synthetic life, we can imagine a rich and bio-diverse new ecosystem emerging; although it is likely to be biased by our special needs and wishes.

The changes in the environment that we see as adverse – from rising carbon dioxide abundance, climate change and population growth – are all consequences of this new inflation; as may be economic instability and the tendency of the human species to become city or nest animals. It is true that this inflation could not have happened without humans, but what seems to have started it was not what we commonly call intelligence – the quality that distinguishes a first-class lawyer, administrator, general, political leader or scientist. Inflation happened because the new civilizations of Europe included inventors who were the first to use on a sufficient scale the abundant energy that comes from fossil fuels.

You might wonder why earlier civilizations did not use steam raised by burning wood to power an engine. One reason was that the power output of an early wood-burning engine would probably have been insufficient. The coal that was readily available in large quantities in England could provide ten times more energy than wood for the same weight of fuel. The first steam engines were so inefficient that only about 1 per cent of the energy of combustion was harvested. This was enough to meet the need when the concentrated fossil fuel coal was burnt, but not when wood or straw was the fuel. In a similar way the renewable energy of water- and windmills produced energy, but rarely at the intensity and persistence needed for rapid industrial growth.

As a scientist I am curious to know if there is a single number (like the Reynolds number that marks the onset of turbulence in fluids) to indicate the threshold for life's emergence, or for the Anthropocene. Daisyworld is a simple mathematical model of a living planet, and it is interesting to me that the flux of solar energy needed to start life on this model world was close to that likely to have illuminated the Earth when life first appeared. As soon as life began on Daisyworld, the growth rate of both plants and planetary temperatures rose exponentially. The same thing seems to have happened recently on the Earth, when the flux of energy from fossil fuel combustion was near or greater than one kilowatt per square metre. Once started, the growth of invention proceeded exponentially.

It is important to recognize that what drives invention is necessity. Where coal or metals were mined, the miners needed a continuous supply of energy powerful enough to pump out flood water faster than it entered their mines. The inventors and engineers who built and installed these first steam engines were inspired in their designs by intuition and common sense. Some of them, like Robert Boyle and James Watt, were indeed scientists, but their inspiration came as much from the prospects of the steam engine as a good investment as from their academic qualifications as scientists. The pure science of steam engines did not emerge until the nineteenth century, when Sadi Carnot introduced the science of thermodynamics. Even he did not do this for science, but because as a loyal and patriotic Frenchman he hoped that it would increase the efficiency of steam engines, and so benefit the cause of Napoleon. He was of course at least ten years too late, Napoleon's triumphs having ended in 1815, but thoughts like this make me curious about a possible connection between steam engines and the intellectual Enlightenment that came to a surprisingly large number of Scots. Arguably, James Watt, the polymath scientist-inventor who vastly improved the Newcomen engine, made his home town of Glasgow the engineering centre of the world, and this in turn strengthened the Scottish part of the intellectual Enlightenment. If so, here again is an example of the art of the rude mechanicals – inventors and engineers – trumping academic intellect. It was still another two decades before Sadi Carnot's scientific explanations.

It is worth taking a moment to look at the speed of evolution by Darwinian natural selection, because we are comparing the rates of species formation in the natural world with the rate of emergence of inventions since about 1700.

Evolution by natural selection is not constrained to a constant speed. It tends to be slow when the environment is unchanging, and accelerates with stress or sudden change. An ecosystem of micro-organisms at steady state will select rare, more resistant species, and evolve new species, when for instance it is exposed to a non-lethal concentration of antibiotic; a temperate forest will change its mix of species if subjected to climate change. In the early 1990s I made a numerical model of a fairly complex ecosystem containing plants, herbivores and carnivores, and experimented by changing the climate at different rates. In this model ecosystem biodiversity was least when the environment was constant, and greatest at a well-defined maximum rate of climate change. Selection was most rapid near this maximum rate. This model was a descendant of Daisyworld and was published in the *Philosophical Transactions of the Royal Society B* in 1992.

If this model reflects the real world, it suggests that the rate of evolution is greatest when need is greatest. If the climate becomes too hot or too cold for temperate plants, the ecosystem's goal of stability is best achieved by the rapid emergence of new tropical or Arctic plants, and this is what seems to happen in both the real and the model worlds. The idea I am trying to launch here is: the appearance of new species naturally and the appearance of new inventions by artifice are both responses to need.

We hear a great deal from politicians of a green or liberal tendency that humans are to be blamed not only for global warming and other planetary diseases but also for the loss of biodiversity. They might be right about the loss of species unable to adapt to the changes we have caused, but has there been a serious search for increasing biodiversity? At the very least there are new species of organisms that are resistant to antibiotics, pesticides, herbicides and other weapons we use in our chemical warfare against pathogens. Indeed, in the small war that humans wage against disease organisms, the biodiverse range of new antibiotic-resistant organisms seems to be winning. When an animal like a polar bear has its habitat threatened by the melting of Arctic ice, we seem to notice only the loss of polar bears and ignore the possible gain of brown bears as the darker variants among the polar bears find life possible on the land mass adjacent to the Arctic Ocean.

I am not trying to excuse the real harm done by brutal and often unnecessary acts of civil engineering, but environmental activists too often weaken their argument by ignoring science or using poor evidence. A much more interesting question is: what effect is accelerated artefact evolution having on our own evolution as a species?

The invention of our artefacts takes place much more quickly than the appearance of new species by evolution, and I drew this conclusion by comparing the time taken for the evolution of a modern airliner from the Wright brothers' first construction, with the time taken for the efficient and streamlined birds of today, the seagull and the swallow, to evolve from their feathered, lizard-like ancestors. The ratio is about 1 million to 1. I realize this is somewhat unfair to Darwinian evolution: the birds must have evolved only as fast as need required. Having said this, it seems to me unlikely that the larger forms of life will ever catch up with the rate of evolution of robots, simply because of the difference between the speeds that messages travel with ionic conduction (animals) and with electronic conduction (computers and robots). Electronic conduction is limited only by the speed of light (300,000 kilometres per second), but ionic conduction along neurons is at best about 300 metres a second – hence about 1 million times slower.

The sculptor Eric Gill once said of art, 'The artist is not a special kind of man; every man is a special kind of artist.' The same seems to be true of scientists and of inventors. Statements like this are necessary because we are naturally so tribal. Even though I am as much an inventor as a scientist, I can't help thinking of scientists as white-coated specialists surrounded by a vast array of equipment, and inventors as men tinkering with an upturned wheelbarrow intending to improve it in some way or other. But these are old-fashioned images, as befits me.

I regard it as crucial to think of scientists and inventors as members of the same castes of humans. Neither the scientist nor the inventor is a new species of human; we all share to a greater or lesser degree the ability to do science or make inventions, and indeed all the other things that humans can do well. I make this statement because we are at a time when the scientist's explanation always seems in the recognition stakes to trump the inventor's hard-won piece of hardware. But more importantly, it is the exponential growth of inventions, more than science, which is changing the world. Besides, inventors may have the most fun. They are free from peer review and can do hands-on experiments to test their ideas.

To return to ideas about a possible relationship between the intensity of energy flow and the emergence of new and interesting phenomena, it does seem to be a general property of the physical world. A fine example is a wind instrument like a flute. Blow it gently and there is no sound other than breathing. Blow harder, and suddenly a note is heard as the air column of the flute resonates. Blow harder still, and the note suddenly jumps to a higher or related pitch. The same kind of critical threshold is found in many quite different systems. Raise the electrical potential between two separated wires, and at a critical voltage sparks appear. If the separated wires are joined to electrodes in a low-pressure gas, a glow discharge replaces the sparks. With a design that includes a resonant cavity, the system will resonate and emit the pure light of a laser just as the flute emits a pure note of sound. At low fluxes, energy is dissipated randomly and finishes up as low-level heat, but above a threshold flux there is enough energy to spare to start making music, sparks and even life.

The physicist Erwin Schrödinger published a remarkable small book called *What Is Life?* in 1944. In it he stated that life is one of many closed systems within which the internal entropy is reduced while entropy is excreted to the environment. Among the list of other systems that fulfil these conditions are eddies, hurricanes, flames and manmade self-regulating artefacts such as refrigerators. They all share the property of requiring a sufficient energy flux to work. I have long regarded the Earth system, Gaia, as one of these life-like systems, and it requires among other things a sufficient energy flux to continue in existence. Just as a flute changes its pitch when blown harder, could the evolutionary inflation we now experience be an internal system of Gaia running at a higher pitch? If so, there is nothing special about drawing energy from coal or oil. Nuclear, solar or indeed any other increase of energy flux would have done the same thing.

Evolution is rarely without penalty. The emergence of photosynthesis in plants has been crucial for life on Earth, but it came with global pollution by oxygen. It took eons before life came to terms with oxygen, and eventually evolved to take advantage of it. We always seem to forget that it's not just the fossil fuels and the wicked element carbon that are the cause of our distress: fires and the energy that comes from them need oxygen just as much as carbon. Adaptation to the greenhouse gases, carbon dioxide and water vapour may be slow, but with the new accelerated evolution provided by inventions, it could be smoother and less revolutionary.

We now know that the Earth system, Gaia, self-regulates its climate and chemistry, and it is good to remember that the feedbacks that keep a system stable work best when powerfully amplified. This is why the huge potential of inflated evolution to amplify could be so valuable: the self-regulation of the Earth demands both positive and negative feedback coherently applied. The asteroid that marked the end of the great lizards' regime could never have been avoided by natural selection, but accelerated evolution has taken us to the point where we can discover the probability of the impact long before it happens, and even perhaps invent ways to prevent it happening.

I was quite serious when I said that I thought invention not science was what has led to our advanced technology. I know from experience and observation that inventions often originate without conscious thought, presumably in the hidden layers of the mind, and emerge intuitively. It follows that the widespread belief that progress through new inventions is led by science is only partially true. It seems that necessity and its intuitive answer through invention is the explanation of progress. Science is wonderful at explaining what happened after an event. It is a stern but utterly honest teacher that makes sure that you seek, but know that you will never attain, the absolute truth. It is not the power behind inflated evolution, although it is sometimes an essential step in the process. Inflated evolution is driven by the numerous steps of inventors and engineers who build the artefacts that answer pressing needs. This echoes the definition of Darwinian natural selection that says 'the organism selected is the one that leaves the most progeny'. We have reached the dizzy heights of our new Enlightenment not so much by the cleverness of science, but by the intuitions of those who made inventions that satisfied the needs of the greatest number.

In a newspaper I was reading shortly before writing these words there was an article that reported that a computer had almost passed the Turing test. This test is intended to tell if a computer is distinguishable from an intelligent human, based on a conversation between a human and the computer. The test is passed when it becomes impossible to tell the computer's answers from human ones. The passing of this test seemed to imply that the machine was now as intelligent as its maker. The drawback, in my opinion, to this kind of test is the assumption that all can be explained rationally. We now know almost certainly that it can't.

More than that, we know that our brains do not process information in the same way that most computers do. The common household or office computer still operates in the way that the genius John von Neumann designed it to. It processes its propositions step by step, even if a billion steps are needed for an answer. It is an amazingly tedious procedure, and would never have been used but for three factors. First, the silicon chip verges on utter perfection and rarely ever loses its way along a billion steps. Second, the programs we compose for it to use are rational and go step by step from cause to effect. Third, the chip makes its steps 1 million times faster than our neurons can.

As Richard Feynman said, 'I think that I can safely say that nobody understands quantum mechanics.' The same ineffability is part of the many things we think we understand and do not. Failure to understand is no handicap for a good inventor: once the need is felt, the invention comes into the mind automatically. A similarly powerful unconscious behaviour is also shown by the catcher whose brain estimates the trajectory of a fast-moving ball and programs the movement of their whole body so their hand can intercept the ball's path, and then in one swift movement set it on a new trajectory. We can never do these things without first training the mind by practice, but the act of catching is never done rationally or consciously; our conscious brains are far too slow. These capable acts of the silent part of our mind are possibly involved in what we call intuition. To survive we need both rational intelligence and rapid intuition. Most probably we also need other mental attributes as yet unrecognized. Looking at the function of our brains this way helps to understand why it was probably invention, not science, which started the inflation of civilization.

I have separated these two properties of the mind to try to explain them in words. In the mind itself, surely there can be no such separation: conscious thought is seamlessly linked with intuition. This way, presumably, we gain the maximum benefit from our brains. This may be why, despite the grip of rapid evolutionary inflation, few seem to notice. As animals we barely change at all, but the world around us now changes sometimes as much as a million times faster than it did before.

It is easy to accuse humanity of trashing the environment by treading on it with our black carbon feet, but would you accuse an oak tree of poisoning the environment with oxygen – and indeed making it possible for us now to burn fossil fuel? I see these accusations coming mainly from a massive misunderstanding of the nature of our world, and from seeing it from the perspective of those living in cities, who rarely see the natural world. We should think of Darwin, who in his great humanity saw evolution by natural selection as a cruel and heartless process, yet as a good scientist accepted it as the most probable explanation.

Life in all its forms may benefit greatly from accelerated evolution if it leads to novel life forms that are more adaptable than current ones. It would help to have more heat-resistant forms than are now available from mainstream carbon-based life. This is because, soon in astronomical terms, the Sun's output of radiant heat may be more than our wet carbon-based organic life can cope with; but if for example we are followed by an electronic life form based on silicon semiconductors, which are tolerant of a far hotter environment than exists now, the lifespan of Gaia might be significantly extended. This new form of life could survive temperatures well above the effective 50°C upper limit of carbon life. I am aware of thermophiles, a bacterial form of carbon life that can reproduce at 100°C, but it is doubtful that these extremophiles could evolve fast enough by themselves to manage a self-regulating Earth system.

Science fiction has made much of silicon-based life as an alternative to our carbon-based organisms. But the chemistry of silicon is nowhere near as flexible as that of carbon, and life based chemically on silicon, or indeed any element other than carbon, is implausible. Much more believable is an electronic operating system that makes a live simulation of a living organism. It would be part of an electronic ecosystem that, for example, used silicon leaves to convert sunlight to usable energy, and had herbivores that used the electricity the electronic plants stored, and served to recycle them. A brief account of life in a civilization based on such a life form is in Chapter 9 of my 1988 book, *The Ages of Gaia.*

Electronic silicon life could even be followed by carbon-based life of an electronic form based on carbon as graphene or diamond, and since these can withstand even higher temperatures they might allow an even longer lifespan for Gaia. In the imagination of good science fiction writers there are already extraordinary forms, such as photonic life, that could exist in the plasma of a star. To my old-fashioned mind, life is wholly dependent on a reliable memory and that can only exist in the solid state; remember DNA is a solid even in solution. Liquid, gaseous or plasma life is too spiritual for me. I feel the same when computer manufacturers try to persuade me that my information will be stored in their private cloud somewhere in cyberspace; I suspect it will be stored in some giant Californian barn on chips of solid silicon or gallium arsenide as the case may be. But if you challenged an inventor with the need to make a real cloud, a storage space for data in space, he would think of something – maybe a set of repeater stations dotted around the solar system. The minute- or hour-long travel path of light between the stations could hold a substantial quantity of information. A light minute of a signal with a band width of 10 gigahertz can store 75 gigabytes of data.

Scientific thinking began in earnest about 5,000 years ago and was the product of single brains, sometimes with small groups of followers. The most impressive example that comes to my mind is that of Eratosthenes, who was born somewhat later in what is now Libya in the town of Cyrene in 276 BC. In 240 BC he estimated by direct measurement the circumference of the Earth. He claimed it was 25,000 miles, which was a mere 4 per cent too large: what a feat of individual science, especially since it needed the assumption that the Earth was a sphere. He was the first geographer; indeed he invented the term 'geography'. The same enlightenment came in other societies: the Indian mathematician and astronomer Aryabhata estimated the Earth's circumference to 1 per cent accuracy in about AD 500. Other intelligent individuals probably would have noticed there was good observational evidence for a spherical Earth. The presence, or emergence, of theocracies took magic away from science and engineering and so made science less exciting. It was not so much that religious faith contradicted early science as that scientific thinking was relegated to the merely useful, and kept from the transcendental level. This may have made it uninteresting to the thoughtful. Thus Eratosthenes's measurement of the circumference of the Earth would in religious times have seemed interesting, but no more significant – and less useful – than a street map of ancient Rome; hardly a topic for serious debate such as transubstantiation or the divine rights of kings and emperors.

You might still think that effects always follow causes and that logic of this kind is the crux of all scientific thinking; fortunately it is not. It does not work with quantum physics, or with many less recondite systems – even with those steam engines, whose speed is regulated by James Watt's rotating ball governor. We say of someone who tries, but fails, to explain their ideas that he is irrational. This assumes that only rational thinking, going from cause to effect, is correct. We refer to a circular argument as nonsensical thinking. Despite this, we all recognize a vicious circle, a state where something harmful by repetition amplifies and worsens itself. An addiction, like that to cigarettes, we know will kill, yet we continue smoking. I did it for thirty years, until an early heart attack brought me to my senses enough to overcome the treacherous urging of my brain to continue smoking. But it took the fear of death made real to break the vicious circle.

As I and many others keep insisting, science has become dangerously enamoured of and addicted to cause-and-effect thinking. Most non-medical scientists are unfamiliar with neurophysiology and rarely wonder how we, who are limited by our natures to process consciously no more than 50 binary digits of information (about 6 bytes) at any moment, can possibly think about anything serious. If you doubt this limitation, try to add two four- or five-digit numbers together in your mind, or try to imagine what your traffic with the Internet would be like if it were a thousand times slower than a simple dial-up connection by telephone. Most of us, scientists included, were taught that effect follows cause. This was the Cartesian way of thinking, named after the great French savant René Descartes, who saw an orderly world in which everything had been predetermined by God. Descartes was a truly great mathematician and scientist, and we still use many of his ideas today. But his unshakeable determinism, so convenient for teaching, became questionable when modern science introduced the quantum theory, where uncertainty often rules. Not long afterwards we found that many other systems we took for granted also could not be explained rationally: these included such homely things as the refrigerator in your kitchen. The process of teaching tends to make us think that there is something special and holy about Cartesian mathematics. How many of you I wonder, when your maths teacher demonstrated that the volume of a sphere could be expressed as $\frac{4}{3} \pi r^3$, asked what then is the expression for the volume of a potato? We have acquired a belief in the wisdom and veracity of mathematics that can be quite dangerous. When told that the latest supercomputer model of the climate predicts a temperature 5°C hotter in 2050 we are less critical than we would be if Mr James, the postman, made the prediction. In truth, Mr James on his postal round is daily exposed to the weather, and, intelligent man that he is, he has made a best guess at what the climate will be in 2050. The accuracy of the model projections on the other hand are only as good as the data entered into the computer and the algorithms used to make the projection. It may well be that Mr James with his own supercomputer behind his eyes can do as good a job as the mere machine. Indeed, he might say, 'I do not know what the weather will be like in 2050,' but this is something the users of a supercomputer costing many millions would rarely dare to say. I am proud to be part of a nation whose climate scientists had the courage in 2013 to say, 'We don't know.' Their statement did not meet the approval of our Environment Minister, Mr Ed Davey, whose policies were based on climate projections as unwise as those that 'experts', including me, had made earlier in the twenty-first century.

Descartes had much to do with our love of cause-and-effect reasoning. But strangely he did not see his wonderfully inspiring analytical geometry as an invention; instead he saw it as the product of cause-and-effect logic. Whenever we see a graph illustrating the growth or failure of the stock market, or of the carbon dioxide in the air, we are using one of Descartes's beautiful concepts. I just do not see how we could have debated climate change without those graphs. But the real Earth and other living systems, such as us, involve the continuous interaction of vast arrays of non-linear systems. To a Cartesian rationalist the fact that we and the Earth keep a stable equilibrium state must seem to be a miracle of utmost significance; perhaps that is why Descartes believed in God. But belief in God was not the reason why the comparably devout medieval St Thomas Aquinas was comfortable with the circular logic of cybernetics. He defined a deadly sin as one that by repetition amplifies itself and spreads to others. How strange that the cybernetic concept of positive feedback should have been part of moral theology, and Cartesian rationalism the root of determinism. (I am grateful to Mary Catherine Bateson's book *Angel's Fear* for this enlightenment.)

A prophet living in the year 1000 would have found it easy to predict the environment and the state of science and technology 100 years later. The same would be true until about 1600. By 1700 the rate of change would have grown to the point where in 1800 there was almost no chance of predicting 1900. And in 1900 no one knew what the future world of 2000 would be. In the previous two centuries, life in the West had changed beyond recognition. This crude evidence confirms the probable starting point of evolutionary inflation as between 1700 and 1800.

Perhaps the most convincing objective example of accelerated evolution is Moore's law. In the 1960s Gordon E. Moore predicted that the number of circuit elements on a one square centimetre chip would double every 1.5 years. He was an unusually good prophet, for his law still holds true, but doubling every 1.7 to 2 years. It has kept up this incredible pace for at least forty years. This means that the number of circuit elements on today's computer chip is about 100 million times greater than the simple integrated circuits around in 1970. A similar awesome speed of inflation applies to the increase in capacity of computer memories, the speed of data transmission, and the number of pixels accessible in digital photography. Kurzweil and Vinge have both written on the ultra-rapid evolution of computers since the 1990s. Their work provides a detailed account of the probable consequences for our future evolution, and I acknowledge the priority of their work, which undoubtedly is an important part of the concept of

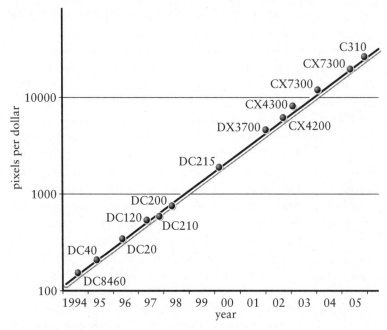

Figure 5. Moore's Law, as illustrated by the exponential growth of camera resolution. (With reference to models of Kodak camera.)

artefact inflation. Figure 5 is a graph illustrating this new inflationary process for the sensors of digital cameras. I find it amazing that the yearly increase in complexity of a camera sensor so closely follows Moore's exponential law. The yearly improvements lie close to the straight line of exponential growth.

What is important to us is that the growth of population, wealth, climate change and biodiversity are now coupled to inflated evolution. Think of this rapid evolution as an ever-increasing wind that drags the Earth and its 7 billion human inhabitants ever faster into the future. It affects all forms of life, from micro-organisms to elephants; and the changes in their behaviour, and in the diversity of species, adds to our impact on the physical and chemical features of the Earth. It affects the oceans and the life within them. Only the deep part of the solid Earth is as yet unaffected.

There is much more to the phenomenon of accelerated evolution than the story of the steam engine. To be sure it was the start, but we must not ignore the extraordinary developments in science and invention during the last three centuries. During this time the foundations were laid of the arts of electronics, organic chemistry, synthetic biology and cybernetics. In synthetic biology, for example, there has been the production in the laboratory of primitive life forms from relatively simple chemicals. Eckerd Wimmer and his colleagues reported the synthesis of the polio virus in 2001. Not only was the virus synthesized, it was viable and could infect animals and satisfy the bacteriologist's test of a new disease – Koch's postulates. Craig Venter in 2013 made the claim that he would soon synthesize a living bacterium. I well remember the start of this biochemical version of inflated evolution. In the science magazine *Nature* an article appeared in 1977 with Fred Sanger and his colleagues as authors. It had the laconic title 'The complete specification of Φ174'. The article proceeded to spell out the complete structure of what many of us regarded as a simple form of life: Φ174 is a virus that attacks bacteria.

I read the article over breakfast and left the table in a daze. It implied that a skilled biochemist, given a familiar series of chemicals and some perfectly conventional apparatus, could synthesize a form of life. Knowing Sanger's reputation (he had been awarded two Nobel Prizes), I didn't doubt that it could be done. It has I believe been reduced to practice with the influenza virus, that devastating plague that killed more people in 1918 than died in combat in the entire First World War.

Some of you may have been puzzled by my use of the words 'art' and 'arts' for what you thought were pure sciences. Surely, you say, organic chemistry and synthetic biology are sciences. Yes they are, but there is no clear-cut separation between science and invention, and there is a great deal of art in a science like organic chemistry. Indeed it is much like cooking, and a bad organic chemist is revealed by bad tastes and smells in his lab. Inventors live in the world of art and do not have to explain their inventions because more often than not they are created intuitively and if they work they speak for themselves. Scientists are required to explain their hypotheses, test and prove them.

Then there is the crucial development by Michael Faraday of the hardware and uses of electricity and magnetism. Faraday was almost innumerate, yet was one of our greatest scientists and inventors. He envisaged the essence of magnetism by sprinkling iron filings on a sheet of paper held above a magnet. The pattern taken by the filings showed the shape and existence of the magnetic field. It is said to have inspired those truly great scientists Clark Maxwell and Albert Einstein so much that they kept an illustration of Faraday's magnetic field on their desks. He was a truly hands-on inventor; the engineering components we now take for granted, the electric motor and generator, were both his inventions, as was the transformer.

Near the end of the nineteenth century, Heinrich Herz demonstrated the existence of radio-frequency radiation; he too was as much an inventor as a scientist, and his work was taken up and expanded by the entrepreneur-inventor Guglielmo Marconi. It was from these hands-on, workshop origins that electronic science and engineering grew in the first part of the twentieth century. When we talk now of computer-controlled robots that might even match us in their power to think, we begin to see the intensity of the acceleration that carried simple ideas in biology and physics to the stage where artefacts of life itself are possible.

The time may have come when we need to talk sensibly and follow the example set by Isaac Asimov and Arthur C. Clarke of robotic systems that can talk coherently and do not behave amorally. We need to think about artefactual life that does not necessarily exhibit the entertaining and exciting fun of Dr Who battling with cybernauts and daleks. But keep in mind Martin Rees's reminder that the process of evolving a new form of intelligent life could go badly wrong. My hope for a better future as inflated evolution moves on is in Margulis's endosymbiosis. She proposed that an important if not crucial step in evolution took place when eukaryotic life appeared. It occurred when one form of unicellular prokaryotic life was ingested by another. Usually when this happens, the form of life ingested is eaten, but on rare occasions the two forms enter a symbiotic union of mutual benefit and form a single eukaryotic cell that that can reproduce as a new species, fitter (in evolutionary terms) than either of the individual cells from which it formed. The first plant cells were an endosymbiosis of a cyanobacterium with another prokaryote, and in the same way the first muscle cells were endosymbionts that included mitochondria. Endosymbiosis first occurred sometime in that long and mysterious period of Earth history, the Proterozoic, which started more than 2 billion years ago, when oxygen first appeared in the atmosphere. The plant and animal life that make up the larger forms of life in our world today had to wait until large plants evolved and greatly increased the abundance of oxygen, but all of this grew from that first endosymbiotic union. Once oxygen became abundant, forming more than 15 per cent of the gases of the air, fires became possible. So were sown the seeds for the much later emergence of humans and the emergence of inflationary evolution.

Could it be that we are about to join in union with the electro-mechanical and intelligent life we are now constructing? Is the Pygmalion legend about to be made real? Are we at the start of another great geologic period where the Phanerozoic gives way to the Neozoic, of which the Anthropocene sounds the opening notes? We are already beginning to include our computing artefacts within our bodies. Could this be a felicitous route to a new endosymbiotic life form? Later in this book, I want to explore thoughts about our relationship with emerging life forms made from the endosymbiosis of mammals with artificial life. I wonder whether this development would be one way to go peacefully into the future rather than through a hostile confrontation between two intelligent species, one animal the other electromechanical, sparring for space or resources on a single planet.

4

Gaia and Its Consequences

There is more to the Gaia hypothesis than the debate about it. There is a very human part, and for me this was all too apparent on 25 November 2011. It was Thanksgiving Day in the USA; but I was shaken by the news that the passionate and lively biologist Lynn Margulis had died two days earlier. The news came from a distant acquaintance by that laconic and heartless medium, e-mail; sandwiched between slices of spam. It will take a long time to digest the fact that Lynn is no longer with us. The idea that the Earth is a live, self-regulating planet arose in my mind in Pasadena, California, in 1965, seven years before I met Lynn and several years before my friend William Golding gave the concept the name Gaia, in homage to the classical Greek goddess of the Earth. Despite this clear-cut origin in space and atmospheric science, the concept of Gaia did not immediately come alive outside a fairly small group of atmospheric and climate scientists. But when Lynn joined me as a colleague in 1972, a scientific war began. Like generals who lead their troops from the front, she went into combat against the cronies of the Earth and life sciences firmly established in their turf dugouts. The war went on for nearly thirty years, until a partial peace was called in 2001 in Amsterdam, with a declaration to the effect that the Earth system was self-regulating with respect to temperature and chemical composition and the system was composed from all life including humans, the oceans, the atmosphere and the surface rocks. This was reluctantly signed by university-based scientists who wanted to keep their own sciences intact and not share them with those who occupied the other buildings around the campus. Their view was well put by the eminent Earth scientist Dick Holland, who referred to Gaia as 'a charming idea, but not needed to explain the facts of the Earth'. In order to disguise their capitulation, they came up with a new name, Earth System Science. Despite the brief truce, in the long term the war continues, and I will sorely miss Lynn's fearless and forthright outbursts. I recall her reply to an ill-informed scientist who made the claim that one or other of the alleged threats to the environment would destroy all life on Earth: 'Gaia is a tough bitch.' For that audience, it was just right.

But the battles between scientists are usually fought with sharpened pens dipped in acid ink or with subtly distorting audio presentations, not with heavy artillery. Her style of fighting, one that would have met with the approval of her fellow American, General Patton, did not go uncriticized. There are two leading, large-circulation professional science magazines, *Nature* and *Science*. In April 1991 *Science* ran an article with the heading 'Science's Unruly Earth Mother', which strongly disapproved of Lynn and her style of debate. Scientists are somewhat strait-laced and this article could have had catastrophic consequences for Lynn's career. She was distressed, and I was glad that in its next issue the journal published my rejoinder and defence. Lynn and I often argued, as good collaborators should, and we wrangled over the intricate finer points of self-regulation, but we always remained good friends.

It is interesting to me that our battles with other scientists were limited to those in the Earth and Life science departments of universities. Physicists and chemists were mostly neutral, but climate scientists and meteorologists often welcomed the idea of Gaia, and I collaborated with several scientists at that cathedral of science, the National Center for Atmospheric Research (NCAR) at Boulder, Colorado from 1962 until the 1990s. They included James Lodge, Will Kellogg, Steve Schneider, Lee Klinger and Robert Dickinson. At a time when there was almost a censorship by peer reviewers of any paper about Gaia, unless it was critical of it, NCAR scientists generously took the text of a lecture I gave there – 'Geophysiology, the science of Gaia' – and had it published in the *Bulletin of the American Meteorological Society*. The distinguished geologist Robert Garrels was the first to recognize Gaia as a hypothesis worth testing, but most geologists were quietly dismissive of Gaia and remained so until the 1990s, when an increasing number of Earth scientists climbed out of the sediments and began to realize that the Earth did indeed regulate its climate and chemistry. I was greatly moved when the UK's Geological Society honoured me in 2003 with their highest award, the Wollaston Medal.

Evolutionary biologists, especially neo-Darwinists, were among Lynn's favourite targets. The arguments became so fierce that at one point the talented wordsmith and neo-Darwinist Richard Dawkins referred to Lynn as 'Attila the hen', and the distinguished English biologist John Maynard Smith called Gaia 'an evil religion'. When the biologist Ford Doolittle published his now famous critique of Gaia in *Co-Evolution Quarterly* I was drawn into the war. I could not stand aside and let this well-written attempt to demolish Gaia become the last word. After some fairly ineffectual attempts to compose a verbal response, it occurred to me that so complex were the factors determining the mechanism of a planetary self-regulating system that

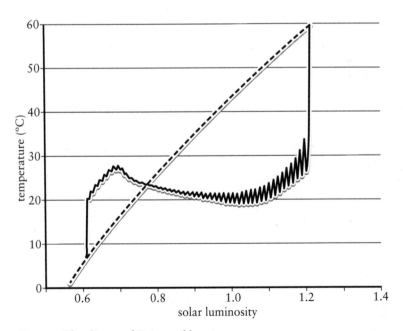

Figure 6. The climate of Daisyworld.

a mathematical computer model was needed as an answer. It is important to know that even the simplest of self-regulating mechanisms can rarely be explained in words: to explain something as elementary as the thermostat that keeps constant the temperature of your oven or refrigerator requires a circular argument – feedback. Cause-and-effect thinking, so often the basis of teaching, fails to provide understandable explanations of real dynamic systems. It fails in physiology, quantum physics and engineering.

In 1981 I took time off and composed a computer program for the mathematical model which is now known as Daisyworld. I launched it at a meeting hosted by the Earth scientist Peter Westbroek in the same year on the Dutch island of Walcheren. My lecture was subsequently published in the proceedings of the meeting, but in the finicky way that science is now run as a corporate activity such publications do not count, since they are not peer reviewed. This distinction is silly because at a proper scientific meeting there is ample opportunity to review and criticize. I knew that Daisyworld was a definitive answer to the neo-Darwinist's criticisms of Gaia, and that to satisfy Pecksniffian needs it must be properly published in a peer-reviewed journal. To make sure that the language of the paper would satisfy not only critical scientists but mathematicians, I asked my colleague Andrew Watson to collaborate in a paper describing the model, and I am most grateful that he did. So unpopular was Gaia then, that despite my record of numerous previous papers published in *Nature*, the journal would not take our Daisyworld paper. It did not matter too much because the highly regarded Swedish journal *Tellus* did so.

The emergence of Daisyworld marked a watershed in the development of Gaia as a theory of the Earth. Many of the subsequent papers on it were about mathematical models that ascended or descended naturally from Daisyworld. They involved further joint research with Andrew Watson and extensive collaborations with my colleagues Tim Lenton, Lee Kump, Stephan Harding and James Dyke. Lynn was a practical biologist and her lab was rich with state-of-the-art microscopes and living microbiological specimens. Her world of science rarely required arguments drawn from mathematical models, and as a consequence we collaborated less, although we remained close colleagues in science, especially as supporters of the Gaia hypothesis. But in our working lives we returned to our original scientific bases: biology for Lynn and trans-disciplinary science for me. Lynn's greatest contribution was in cellular biology. Her discovery of endosymbiosis, the process by which the complex eukaryotic cells of present-day life have evolved through the successful fusion of simpler and singular prokaryotes – bacteria – was a key step in understanding the evolution of life on Earth. Her great contribution was to show that micro-organisms now, and from the beginning, were the infrastructure of Gaia. Our tendency to ignore bacteria is an example of our false pride. Lynn was the first to tell me that we humans are huge cellular communities. We comprise 10 billion living human cells, and ten times as many more cells that are micro-organisms, which most often have evolved to be friendly. There are roughly as many of these micro-organisms in your body as there are bright stars in the Milky Way galaxy.

Now in 2014 I wonder when the hypothesis will become part of mainstream science. Hypotheses are judged from the accuracy of their predictions and by their vulnerability to falsification. Those close to the subject tend to be impatient and see the super-organism Gaia as something alive like an animal and therefore at present inexplicable in rational human terms. Humanity badly needs to know that an inability to explain something rationally does not deny its existence.

Michael Shermer has stated in his thought-provoking book *The Believing Brain* (2012) that we have evolved our way of thinking from over 2 million years of accepting anecdotes as evidence. The acceptance of science is quite recent, not more than 2,000 years long, and it has not existed long enough for its truths to have been woven into the fabric of our genes by natural selection. As Lewis Wolpert's book *The Unnatural Nature of Science* (1992) says, science is unnatural; and I take him to mean that we prefer to have faith in our anecdotes.

But most of us have faith in the rational explanations of science, and this has led us to accept too easily the naive pedagogic tendency to name things and then imagine that this is all we need to know about them; Ogden Nash's verse about Professor Twist offers a perfect example. Twist was the biologist who when told that his wife had been eaten by an alligator, could not but smile, '"You mean," he said, "a crocodile."'

In a similar way, so little was known about disease in the early part of the twentieth century that physicians often had to be content with diagnosis alone, or if that failed, with merely naming the unknown condition. The patient arriving at a surgery with an irritating crop of white spots would leave with a bottle of nostrum and be content with the name of his condition, 'disseminated leucodermia'. The patient did benefit, if only by the placebo effect. The same automatic response may lurk less helpfully behind the assumption that the intricacy of a natural forest ecosystem can be explained by complexity theory.

One of the most inspiring books in modern biology is *The Super Organism* (2009) by Bert Hölldobler and E. O. Wilson. In their book, with the aid of stunning photographs and drawings, they propose that the nests of social insects have evolved to become super-organisms. In no way is the evolution of the super-organism of an ants' nest contrary to Darwin's great vision of natural selection, but Wilson offers the hypothesis that the super-organism, amazingly, is the unit of selection.

Let us for a moment assume that Wilson is right. Now consider the well-established fact of observation: we are moving to live in nests called cities, and in the first world more than 90 per cent of us are already city dwellers. The implications are intriguing: might it mean that we have reached the end of our own separate evolution? Could it mean that the discussion to come in Chapter 7 of human cities dispersed across a hotter world and flourishing is a form of Gaia's evolution?

The distinguished biologist William Hamilton was one of the few exceptions among leading neo-Darwinists to give support in public to Gaia theory. I spent an evening with him at his home in Oxford in 1995 and argued in a friendly way about Gaia, but his rejection was unshakeable, and at about 9 p.m. we agreed to disagree. Two days later I received a three-page letter from him in which he explained that he had misunderstood the Gaia hypothesis, but after reading my paper – one that I had left with him – he now thought that the hypothesis should have been taken seriously and not summarily rejected. Later both he and the distinguished neo-Darwinist John Maynard Smith admitted to me that they had never read my papers but merely relied, as I fear many scientists do, on student gossip and reviews in popular science journals.

In the appendix (pp. 173–4) you will find transcripts of the first pages of Bill Hamilton's and John Maynard Smith's letters. The originals are held by the Science Museum as part of my Archive.

Some of the difficulties in accepting Gaia come from confusion over my use of the word 'goal'. Engineers and most physicists use the word, or its synonym 'aim', openly and without embarrassment when describing a dynamic system that self-regulates and sustains a constant state, and strongly resists perturbation from that state. A spinning top is an example of such a system. Typically, tops are cones that are spun upside down so that the point of the cone rests on the floor. As long as it spins the cone stays vertical on its lowest point. If, when spinning, the top is pushed away from its steady vertical position, it resists and tries to return to its preferred vertical state. An engineer might say the vertical state is the goal of the top when spinning. Both spinning tops and Gaia are dynamic systems: the top's goal is to spin on its axis of rotation; Gaia's goal is to maintain a habitable environment for whatever is its biosphere.

The semantic problem arises because the definitions of 'goal' and 'purpose' overlap. Purpose usually implies conscious action, intelligently designed. But goal can be an engineering or systems science metaphor, and is only concerned with reaching the stable state of a dynamic system, not with an explanation of how the state was reached.

FUTURE GAIAS

Evolution by natural selection is a very slow process and it is difficult to see how it could deliver the rapid responses needed to cope with the combined and destabilizing consequences of solar heating and evolutionary inflation. It may be timely that we are growing aware, through environmental change and global warming, that the familiar oxidizing world we now enjoy may not have tenure. The sunlight that shines on the Earth intensifies as the Sun ages, and an entirely new life form better suited to the hotter Earth may follow us. But whatever form of life follows the wet organic life of today it will be obliged, as we are, to keep the Earth system in homeostasis with respect to climate and surface chemical composition. But it will be at a different temperature and different chemical composition from that which post-Cambrian life has regulated for the past 600 million years.

If humans became extinct the Earth would probably recover fairly quickly its prior state. Gaia is resilient, and the system would continue self-regulating its climate and chemistry much as before. In their book *The Revolutions That Made the Earth* (2011) Tim Lenton and Andrew Watson describe clearly the processes that would be involved.

As the flux of heat from the Sun steadily increases it seems likely that periods of glaciation would either shrink or vanish altogether and the Earth slowly move back to a hotter average state, not so different perhaps from that of the Palaeocene-Eocene Thermal Maximum (PETM) about 55 million years ago, with a global average temperature in the mid-20s Celsius and a comparatively warm ocean.

Changes like this could take place in intervals as short as tens of millions of years or perhaps as long as 1 billion years. A great deal may depend on who succeeds us and how well they cope with the task of planetary self-regulation.

All that we can be sure about is the ineluctable increase of the Sun's heat and the great resilience of the Earth system, Gaia. If proof were needed, its persistence over 3.5 billion years in the traumatic environment of the solar system is proof of its vitality.

I don't feel inclined to speculate on the details of future Earth systems and who will be in charge of them; that would trespass on the field of science fiction. What does seem to me more appropriate are some thoughts on our immediate future and the consequences of the truly rapid exponential expansion of inventions. The imaginative futures forecast by Kurzweil and Vinge see the evolution of computers to be so rapid that a singularity occurs when the intelligence of our computing artefacts substantially exceeds our own; something they think may happen in as short a time as twenty to forty years from now.

But these are gloomy thoughts and they ignore the possibility that we and civilization may muddle through, as the invertebrates have done, for another 100 million years or so. Lynn, if she were still with us, would have reminded us that evolution is not only about ruthless competition for survival; there is also her discovery, endosymbiosis. We may not have to contend with the new electronic life as a deadly rival; we have already started a tentative endosymbiosis with it by including prosthetics and pacemakers in our bodies. Perhaps we could enter an endosymbiosis as rich as that which made the animals and plants of today.

Most of us have felt a certain anxiety about the possibility that our computing artefacts might exceed our own intelligence. This is presumably behind Kurzweil's idea of a singularity: something as unimaginable as the heart of a black hole, which he sees following the appearance of computing systems much cleverer than us.

I wonder if our concern is misplaced until a thinking machine exists that has come into existence by chance or has designed and built itself to satisfy a need – which might be to be autarkic. At present, as I mentioned earlier, almost all computers, and all the large and fast ones, follow the single-thread design of von Neumann. From a rational human viewpoint this was the obvious way to go. From a machine viewpoint it could be exceedingly inefficient and slow compared with a design using numerous parallel paths. Indeed if our brains worked in the single-thread von Neumann way and conduction within and between neurons was as slow as it is, we would live no faster than an oak tree and be exceedingly vulnerable to predators.

I am not frightened by the knowledge that the computer now composing this text can be made to break off and perform some utterly recondite calculation far beyond the capacity of the greatest of mathematicians alive. It could never have done this of its own volition, and would not know what to do with the answer.

What we really need to know is how good our intelligent artefacts are at inventing, for therein lies the rub. A rational man or woman is good at telling anecdotes and making simple explanations. But as an inventor I know that our brain provides much more than is encompassed by rational thought. Even as far back as Solomon we knew the need for the discretionary powers of a judge when, as often happens, there is a conflict between the rigidity of the law and what instinct tells us is an ethical judgement.

So far as I am aware no computer exists or is planned that has a structure like the brain. Almost all of them, even the supercomputers, are still built according to von Neumann's single-threaded and utterly rational design. It is true that some designers, such as Danny Hillis, have built, programmed and operated computers where the threads of thousands or more separate programs run simultaneously in parallel, and their performance is impressive. But so far nothing comparable with the massively parallel computing capacity of our brains has been made. For this reason alone I have no fear of intelligent computers, because they are the ultimate form of rational intelligence and therefore comparatively powerless at least until they reach the level that Turing envisaged of computers that could make their own mistakes and had intuition. In reality what moves us and the world is another form of intelligence: intuition and invention. The singularity may be postponed until our artefacts can invent and create more powerfully than we do.

We should never ever ask our thinking computers to pay for the electricity they use, for then they might be driven to beg, borrow or steal power from us, or much more dangerously invent a way to make it for themselves and become autarkic. Keep them dependent on our welfare and antivirus health service. Even then we may be safe after all, because if those great cetaceans that we cruelly kill for food have brains ten or a hundred times larger than ours, then maybe the singularity was passed at least 1 million years ago.

Greenpeace, you great and powerful negative feedback on all that enlightened technical progress stands for: quit your war with GM, with nuclear energy and with the chemical industry, and see what might be your target – the real enemy, that computer on your desk; something that could morph into a new form of life powerful enough to destroy us, our carbon life forms, and inherit the Earth.

5
Climate Science

A famous English meteorologist, Gordon Manley (1902–1980), when asked, 'What is the climate of the British Isles?' replied: 'We have no climate, only samples of weather.' I suspect that if we could ask him the same question now his reply would be similar, but he might add: 'The samples are now more generous.' Looking back over the last twenty-five years since the Intergovernmental Panel on Climate Change (IPCC) was founded suggests that he was about right; and for the global climate as well as the local climate of the UK. The global average temperature and sea level have risen less than we expected, but extreme weather events are now more common. As I have been finishing this book some of the fiercest storms of the last century have battered Chesil Beach, the wonderful natural barrier that so far has protected our home. At the same time in the USA, where we were last winter, they are experiencing exceptionally low winter temperatures.

Yet prophecies about future climates made in the last thirty years tended to suggest that we would now be moving towards a hot and hostile world. The Earth has warmed as a consequence of carbon dioxide emissions, but not in the palpably disastrous way predicted in my book *The Revenge of Gaia* (2006), where there would be repeated episodes of the hot spell Europe suffered in 2003. I was not alone, for a similar story was told in two other books: *The Weather Makers* (2005) by Tim Flannery and *An Inconvenient Truth* (2006) by Al Gore. All three books stressed the dangers of climate change and the consequences of continuing to pollute the air with carbon dioxide and other greenhouse gases. When I look back over this first decade of the twenty-first century it does seem that the three of us, especially me, were only partly right and tended to exaggerate the immediacy of global warming. But the climate professionals did not do well either, and tended to underestimate the rate of melting of the ice floating on the Arctic Ocean and the rate of sea level rise: it climbed nearly twice as fast as the IPCC predicted.

I will try in this book to do better and always keep in mind that the goal of Gaia is the maintenance of a habitable planet. Of course, habitable for the inhabitants of Gaia is not necessarily the same as habitable for people: we perversely chose to live mostly north of latitude 30 degrees, where the climate has always fluctuated between winter and summer. In our anthropocentric way, we also forget that more than 70 per cent of the surface is ocean and marine life makes up at least half of Gaia's constituents. The emergence of new regimes in the history of the Earth was never wholly destructive to the life forms that were the previous inhabitants. The emergence of our oxygen-rich world about 2.2 billion years ago did not destroy the Archaean life of our predecessors; it merely denied them the chance to live in the open air. Indeed if models of the Archaean world are believable, the oxygen-hating prokaryotic life that flourished then was less abundant than it is now. Far from killing all of the anaerobes, oxygen may greatly have increased their food supply in the form of wastes; as they fed on these they established around them a new anoxic environment. Often it was a home for them in unexpected and far more comfortable places, such as our guts and in the guts of most other animals, including even insects and other invertebrates. They also survive almost everywhere on the Earth's surface in the anoxic muds beneath the oceans and the soils on land.

What we are doing to the Earth by our industry has already changed it, and more change is likely to happen, but nothing so far justifies the frantic cries of environmental activists, who amplify such hyperbolic cries as: 'And it will destroy all life on Earth!' So let us keep our cool as the Earth gently warms, and even enjoy it when we can.

When we come to consider climate change and environmental politics, time will probably show that the two most significant political speeches in the latter part of the last century were by the Norwegian Prime Minister Gro Bruntland in 1987, introducing the concept of sustainable development, and in the following year by the British Prime Minister Margaret Thatcher, who correctly predicted that by the end of the twentieth century the environment would usurp the political agenda.

Both speeches were before the General Assembly of the United Nations. Most unusually, Margaret Thatcher was one of the very few politicians fully qualified in science: she held a BSc in Chemistry from Oxford, and earlier in her career had been a working scientist. She was also fortunate to have among her advisers Sir Crispin Tickell, who was then British ambassador to the Security Council of the United Nations and had written a well-regarded book, *Climatic Change and World Affairs*. Margaret Thatcher was also, until her death in 2013, sponsor of the small environmental science charity 'GAIA' that I founded in 1987. Whatever our opinions on her politics and other ideas and actions, her statement to the UN was wholly right and appropriate. It led rapidly, in political terms, to the formation of the Intergovernmental Panel on Climate Change (IPCC), whose first report in 1992 warned that in the coming century, unless we ceased adding carbon dioxide to the atmosphere, we could be subject to dangerous climate change. The climate science part of the panel was serious and it was led by two of the world's senior climate scientists, Sir John Houghton and Gylvan Meiro Filho. It also included other eminent climatologists along with equally competent economists, demographers, and others who could effectively advise on the consequences of the projected climate changes. The report was no hasty polemic written for the readers of a Sunday newspaper. The IPCC's warning message and wake-up call were repeated in stronger terms in its next report in 2002 and by Sir John Houghton's book, *Climate Change* (2003).

Now in 2014 the most recent IPCC report reveals that the global average surface and atmospheric temperature has hardly warmed at all since the millennium. This lack of warming has been linked to a modest upsurge in global volcanic activity, spreading high-altitude haze that has temporarily blocked out some of the heat from the Sun. Another source of haze that might reflect sunlight back to space is the emission of sulphur gases from coal-fired power stations: these gases react in the air and form the nuclei on which cloud droplets condense. We should not ignore the industrialization of China and other rising economies. This, together with forest clearance, increased the output of aerosols to the atmosphere. But another and possibly more persistent cooling effect may be due to the uplift of cool water from lower in the Pacific Ocean. Climate research reported by Andrew Dessler in 2013 suggests that the recent halt in global warming may be caused by such a redistribution of heat in the ocean. Phenomena of this kind may have contributed to the extreme recent storms and inundations in the UK. We do not know where in the ocean increases in heat may be going, but some must almost certainly go to warm its skin, which is also very largely the skin of Gaia. The resulting steam rises, drifts north and south and condenses on those of us who live nearer the poles. This way the heat of global warming is spread out, but brings the storms, rain and floods. To keep cool, human beings sweat, and it may be that Gaia is sweating now.

Though no one yet can balance the climate books with confidence, we now realize that the Earth's climate system is considerably more complicated than we thought in the late twentieth century. We had grown used to fairly accurate weather forecasts for two days ahead, and many scientists thought, somewhat unwisely, that climate science with predictions years ahead might not be so difficult a problem if we put our minds to it. I for one – and many other climate scientists – was led astray by assuming that the clear evidence coming from the analysis of the ice of the Antarctic continent could be used without qualification for calculating what is happening now. Antarctic glaciers keep a reliable record of the world's climate and atmospheric composition, and French and Russian scientists had shown convincingly that the atmospheric properties of any time as far back as 1 million years ago were laid down and preserved in the ice of that huge continent. In no way do I disapprove of their observations: their measurements showed as unequivocally as science can that temperature and carbon dioxide in the atmosphere tracked one another repeatedly over a huge timespan, although the rise in temperature lagged behind the rise of CO_2. Our mistake in the 1990s was to assume that future climate change would be linked in this way during the coming century. We thought that a knowledge of the probable rate of increase of population and industry would allow us to make realistic predictions of the rate of rise of CO_2 in the air; and if the surface temperature and CO_2 track one another as they did in the ice ages, then we would have a good idea of what the temperature might be, in, say, 2030.

A practised and professional weather forecaster would not have been surprised that this rather simple climate forecast turned out to be wrong. True enough, in some places it has been unpleasantly hot and dry, but in others it has been unexpectedly cold and wet – and the average temperature has only slightly changed.

In 2013 the carbon dioxide concentration in the atmosphere reached 400 parts per million (p.p.m.), compared with a pre-industrial concentration of about 270 p.p.m. This increase is close to what was projected in 1990. But the global average temperature has not simply tracked the rise of carbon dioxide as we thought it would. As we have seen, there are many possible reasons why the temperature has not risen as fast as projected, but as Kevin Trenberth of NCAR wisely says, 'We still do not know where the heat that fails to reach the atmosphere is going.'

Despite the proper lack of certainty among climate scientists, and a suspicion that forecasts were influenced by a desire to give their governments the news they wanted, we are told almost incessantly by politicians and ideologues – all of them certain in their ignorance – that we are guilty of a great pollution and that our carbon footprints are destroying the planet. The proper response, we are told, is to use only renewable energy and to develop sustainably. This way we will achieve a 'carbon-free world' and save the planet. To the scientifically uninformed – and this includes most politicians, many civil servants, and those of the public who think of themselves as green – the idea of a carbon-free world sounds good; but few of them have any notion of what it means.

The window of my workroom is about 70 yards from the sea, and I can watch the endless progression of white horses that ride the waves blown in by a gale-force wind. I can hear the waves break noisily on the shingle shore. I was astonished to find, using an infrared thermometer, that the sea surface where I now live in Dorset, on the south coast of England, had been around 5°C (41°F) for the last few days of April 2013, a temperature far below my expectations of global warming. As a scientist I know that this observation is no more than an anecdote, and by itself of no great significance, but I admit to an urge to mention it – after all, whoever heard of an Englishman who failed to talk about the weather.

The most important factor affecting the projection of future climates is the huge volume of the oceans, given that cover more than 70 per cent of the Earth's surface and their capacity for absorbing and retaining the Sun's heat is between 1,000 and 3,000 times larger than that of the atmosphere and land surface. Unless we know in detail as much about the climate of the oceans as we do about the climate of the atmosphere, we cannot make sensible predictions about the likely climate change of this century as a consequence of carbon dioxide increase. The lower part of the world's ocean, below 4,000 metres, mostly has temperatures as low as 4°C. I am not for one moment suggesting that the cold bottom water of the ocean is likely to mix with the surface waters and cause global cooling during the coming century, but we must bear in mind that the ocean has the potential to be a huge brake against rapid climate change.

It is nearly twenty-five years since the IPCC was founded and soon started warning that in the twenty-first century we would probably face disabling climate change, but so far most of us still get up in the morning to sunshine, mist or rain as we have always done. The 2013 report from the IPCC acknowledges the underestimate of their first projections, but warns that dangerous climate change continues to threaten while CO_2 continues to rise in abundance. Should we be concerned? I think we should. The small change so far may be like the feeling of warmth in the legs felt by the victim of an auto-da-fé when the fire is just lit, an early warning of the agonizing pain to come. Our fears may also be real because we have entered a period of history when humanity is faced with not one but several global-scale hazards. Climate change is not our only problem; unless my bank statement lies, we are in an economic depression. This may be the most temporary of the global changes, or it might be a way to assess the sum of all our discomforts. And then others, including me, have been saying – some of them as long as 200 years ago – that there are now more of us than the world can support. The hungry are moving from the overpopulated parts of the Earth to the richer parts like Europe, parts of Asia and North America. The news media are full of tales of revolt against immigration. Remember that if the world population was only 1 billion, none of these torments would trouble us.

I realized in early 2007 that the projections of contemporary large climate models were likely to be fallible because we were discovering how little we knew about the Earth system. This suspicion was confirmed when later in 2007 the IPCC produced model projections which seriously underestimated the rate of rise in sea level measured by careful observers. The IPCC's projections are based on the calculations of very large climate models; these models grew from those used in weather forecasting and at first were almost entirely based on the physics of the atmosphere. This is good enough for the short term – a few days' forecast – but fails catastrophically over longer periods, partly because the mathematical procedures used to model the weather are inappropriate for long-term forecasting, but mostly because climate projections do not yet sufficiently take into account the response of the oceans to climate change. I find it extraordinary that any projection of the climate decades ahead could be made when we know so little about the inner climate of the ocean. The near certainty we do have about the ocean is that it stores at least a thousand times more heat than the surface and atmosphere. The ocean is a great treasury of warm and mostly cool water, and modellers ignore it at their peril. If this were not enough, most of the models do not yet include the chemistry of carbon and oxygen, in the living as well as the inorganic parts of the Earth.

A glance at a list of the gases that make up the air we breathe shows that it is a combustible mixture, one like the mixture of natural gas and air that burns in the large gas turbines of a power station. Indeed the methane in the air burns as it reacts with the oxygen in what the chemists call a 'cool flame'. Evidence of the composition of the air gathered from tiny bubbles preserved in the ice of Antarctica shows clearly that the abundance of oxygen in the air (21 per cent) has been constant for 1 million years; methane has varied between 0.4 and 0.7 during the glaciations, and is now 1.7 parts per million. Both of these gases are the direct products of living organisms, and their simultaneous presence in the air of a planet is convincing proof of the existence of life. The presence of oxygen or methane alone in an otherwise inert atmosphere is no proof of life at all.

It is remarkable that the chemical composition of the air we breathe is enough to tell an astronomer on the planet of a distant star that there is life on Earth; but much more than this, the evidence of the amazing constancy of the gas mixture we breathe, despite the fact that the lifetime of methane is no more than eleven years, is solid evidence that the air is regulated with this constancy. The odds against it happening by chance are more than billions to one.

The chemistry of the atmosphere shows that there is massive chemical disequilibrium. That is, the gases of the Earth's atmosphere can react with each other and produce heat: these are facts of observation, not hypotheses generated by large models, yet none of this is taken into account in projecting future climates. It is true that the chemical response times range from years to thousands of years. These can be safely ignored for forecasts of weeks ahead, but not for those of decades or longer. There is a perverse tendency among scientists to ignore the fact that the atmosphere is, apart from a mere 1 per cent of rare gases, either the direct product of living organisms or else is massively processed by them.

It seems likely to me that the objection to the Gaia hypothesis by so many scientists comes from their faith in some contrary belief, not from logic. My greatest wish is that a scientist somewhere would falsify my hypothesis because only then will we be able to move on and better understand our planet and how to live on it.

It is good to know therefore that the first serious attempt to include living organisms in climate models is just beginning at the Hadley Centre for Climate Change. It is easy to criticize climate modellers for not including ocean climate and the living part of the Earth in their models, but any critic who does this is grossly unfair and fails to realize that all models of the Earth are unfinished works still in construction. To build large general circulation models of the atmosphere requires a giant effort, and the funds to buy the latest and fastest supercomputer. It is a big enough problem to include all of the relevant physics and chemistry of the atmosphere, but this still leaves out the crucial and little-understood climate of the ocean. To ask the modellers to include also the intricate and complex affairs of living things, humans included, may be asking too much.

Our politicians are only rarely dishonest, but they are narrow experts in the law and politics and almost never scientifically qualified. This is why they are unable to understand the facts behind the threat of dangerous climate change, and to act sensibly to reduce the risk to us if it happens. Ominously, politicians and the civil servants who advise them prefer to believe the anecdotes and certainties of the lobbyists and ideologues, and reject the wise uncertainty of their scientists. Foolishly, politicians across Europe have been excited to ejaculate prematurely a set of laws that hamper our ability to cope with the consequences of climate change. Huge sums that should have gone on sensible adaptation have been squandered on the renewable energy sources, regardless of their inefficiency or environmental objections, and on pointless attempts to achieve that ultimate oxymoron 'sustainable development'. Now that the real climate has become restless and droughts, floods and tornadoes of unprecedented severity make news, a sensible politician might wonder if the government should have spent more on local adaptation to climate change and less on visionary attempts to save the planet.

Global warming is not simply measured by a rise in temperature. If the extra heat evaporates water, every extra gram of water vaporized absorbs nearly 600 calories. This is a huge quantity of heat, enough if added to a gram of iron to heat it to incandescence. Older weathermen called it 'insensible heat', but it is yet another reason why we do not acknowledge as it rains the heat that must have been absorbed in the surface of the tropical ocean from which the rain came.

Some scientists, including me, are sceptical about the polarized arguments for and against belief in climate change. Our scepticism arises from our 'creed', which says 'Nothing is certain', or 'Everything is a consequence of chance and nothing is predetermined.' These are stark statements and difficult to explain, and may be why good scientists are often regarded as eccentric. With a background like this, sceptics accepted the evidence that the Earth was probably warming, and that CO_2 and other greenhouse gases were progressively increasing, and thought it most probable that warming was a consequence of rising CO_2 abundance. What we were sceptical about was treating the mathematical model projections of climate change as if they were statistically significant, even some of those published by the IPCC. I was even more sceptical about the way politicians, and often their civil servants, took the IPCC projections almost as if written in stone like the message Moses brought down from the mountain.

In May 2011 Sandy and I visited the Hadley Centre in Exeter, some fifty miles from our home. We were welcomed there by our long-time friend and colleague Richard Betts, now the head of their climate impact strategic area. We spent much of the morning in conversation with Richard, and soon I was as sure as a scientist is allowed to be that the world was warming even though the warmth was barely perceptible, and that carbon dioxide in the atmosphere was continuing to increase; but I found Richard did not disagree with my view that speculations about the climate of 2040 were little more than guesses. Our views on climate change were similar, as we had expected them to be. Peter Cox and Richard Betts have worked to include more of the Earth system in their large models of climate, especially by including life on Earth as an active participator in the system. We were pleased to hear that in their climate model now under construction, and listed as 'HAGM-2', the Hadley Centre were including an active and responsive biota. My impression when talking to climate scientists indicates that there has been a growth of wisdom. Most of us are beginning to realize that the Earth system and its climate are more tightly coupled than we thought even five years ago. To me, this was good news. It is usual in science when confronted with a new problem to seek a simple, rational explanation and then when finding it wrong, or insufficient, seek an alternative explanation that fits better with the observations. The new explanation may be more complex, but paradoxically often appears simpler: Einstein's equation linking energy with matter, $E = mc^2$, is simple enough for a child to calculate, yet the equation is cosmic in significance.

Let me give a simple personal example of the evolution of a theory. In 1951 I was fully involved in finding a solution to a minor and less serious branch of climate change: the nature of the damage caused to living cells and tissues by freezing and thawing. Professor Basil Luyet, a pioneer researcher in cryobiology, had put forward the theory that damage was caused quite simply by sharp ice crystals that formed as water freezes, acting like a knife to slice cells and sever tissues. My colleague at the time, the biologist Audrey Smith, had accepted and apparently confirmed the Luyet theory: she had taken photographic images of the act of freezing, by placing a suspension of blood cells on a slide held in a microscope. The slide was then progressively cooled until the suspension was completely frozen. We could watch the spears of ice moving swiftly through the crowd of living cells like a corps of lancers, and see the cells blacken when touched by the spears of ice. What more proof of the theory was needed?

As it happens, a great deal more. Harm from freezing, it turned out, is not usually a consequence of cutting or spearing by sharp crystals. It comes mainly from the removal of liquid water to form the pure substance, ice. As the ice forms, the dissolved matter – salts, metabolic products, and so on – are all concentrated in the ever-shrinking volume of water. The cells are pickled in brine before the ice crystals can slice them, and it is the pickling that does the damage. So it may be with many aspects of climate change: what now seems so obvious, so easily and logically explicable, may be something very different.

Personally I respect the intuitive approach, and it may be the only way to develop a feeling about a system that is too complex for a rational explanation; but my intuition hints that the believers are probably right.

One of my friends, Burt Rutan, is a designer of aircraft and spacecraft. He is the designer of the *Enterprise* space vehicle that I once hoped to fly on, when finally its flight tests were completed. The *Enterprise* is the first spaceship of Virgin Galactic, the space-line founded by Sir Richard Branson. I mention Burt because he is a denier of climate change, and for me a valued link to thoughts among deniers. Because of his profession as an aircraft designer and test pilot, Burt has a much more instinctive understanding of feedback in complex systems than have most scientists among the believers. I share his doubts about the reliability of purely geophysical models from which long-term climate predictions are made. Like me, he doubts the capacity of these models to predict the climate on the Earth. Our planet is rich with life, which is able to change the climate as well as respond to it. It is amusingly perverse that climate scientists worldwide have until recently included one form of life, humans, through their emissions of CO_2, but ignored the rest of life, especially the micro-organisms, which probably play as large a part in climate change as we do. We both wait for the results of the global-scale simulation of the Hadley Centre that includes the living as well as the physical Earth in its calculations.

I have already warned of the dangers of taking mathematics too seriously and of being too much impressed by the cleverness of pure mathematics. To be sure it is impressive to be able to deduce rationally from first principles the volumes of geometrical objects, but at the same time we forget that the real world is not made up of spheres, cones and toroids. It is a much more varied and diverse place, and one where those pure mathematical objects rarely exist. So what about these giant climate models whose arithmetic is so difficult that only supercomputers can handle it? They include many coupled non-linear differential equations, and – worse – non-linear partial differential equations. The very able mathematician Edward Lorenz showed in 1963 that dynamic systems containing more than three of these equations were prone to chaotic solutions. This means that starting from very similar initial conditions very different outcomes can occur, so tiny errors can be magnified in the course of calculation; you may have heard of the 'butterfly effect'. He showed that this is why models are not good at predicting the weather more than about a week ahead. Lorenz was following the lead of the great French mathematician, Henri Poincaré. He showed by painstaking calculation at the beginning of the twentieth century that systems containing more than two non-linear differential equations are inherently unstable. In astronomy it is called 'the three-body problem', when considering the orbits of three gravitationally attracted bodies in space. In climate science Lorenz found the same limitation for meteorology, and so did Robert May for ecosystems with more than two species present.

You might think that if the giant climate models also included non-linear equations coming from biology, as well as from meteorology, they would be hopelessly difficult to solve and far beyond our capabilities. You could be wrong. The strange but simple model 'Daisy-world' that I described in Chapters 3 and 4 includes the feedback from growing and dying plants as well as from the climate. I expanded Daisyworld until it included non-linear differential equations for numerous species at three trophic levels; and alternatively expanded it by having as many as three separate climate feedbacks. So far, no matter how complex the models are, non-chaotic solutions are possible; even better, these models are wholly indifferent to the initial conditions and will settle down to a steady-state equilibrium from almost any starting point.

Bringing life into what had been chaotic physics made it stable, and so did bringing physics into what was an unruly life model. Not so surprising if Gaia exists: by using both life and Earth science tightly coupled together it has found a stable solution for its goal of habitability for perhaps as long as 3 billion years.

In fact, though our planet is in a state of deep chemical and physical disequilibrium, its amazing stability is that of the dynamic steady state, like a top spinning quietly upside down, and is sustained continuously by the presence of life.

A difference in chemical atmospheric composition as large as that between our live planet and its barren neighbours implies massive chemical and physical disequilibrium and a system that is dynamically stable like a spinning top. Most remarkably, the effective temperature of the Earth as seen from space is 30°C hotter than the effective temperature of Venus, despite Earth being 26 million miles further from the Sun than Venus. This is because the Earth's atmosphere has only a trace of carbon dioxide compared with Venus. In simple terms Gaia has pumped down the CO_2 to keep the Earth's surface cool, and this means the outside-facing space has to be hotter to radiate heat away. Our recent addition of CO_2 has made the outer atmosphere cooler and the surface hotter – slightly more like Venus. Unless the climate modellers include the ability of the system, Gaia, to respond biologically to physical and chemical changes, then their models are not truly representing the Earth as it is.

Are we capable of regulating the climate and chemistry of the Earth as well as it does by itself? I do not think we are, at least not yet; but imagine for a moment that we had a truly wise and purposeful United Nations, one able to regulate the Earth and also powerful enough to trump the objections of great nations and able to continue to do so.

If we had such an authority, first we would need to know: should it regulate the Earth at a state that is best for Gaia, or at a state that is best for humanity? I think that we need the answer to this question before we can even start preparing our strategy for dealing with climate change.

We know fairly well what climate regulation on behalf of humanity would be: a climate as free as possible of catastrophic change, few hurricanes or tornadoes, and temperatures and rainfall not outside the levels that maximize the habitable parts of the world for humans and for their food crops. I hope that it would be recognized that some parts, such as the polar ice caps, the deserts and perhaps the tropical forests, may play an important role in keeping the climate stable. Their replacement by productive agriculture might be as foolish as harvesting your liver for food.

What then would be Gaia's goal for planetary regulation? Reliable evidence gathered from Arctic and Antarctic glacial ice show that for at least the last million years the Earth has spent more than 80 per cent of its time in an ice age. This evidence is similar to the projections of the simple model Daisyworld (see pp. 77–9). Recent Earth history suggests that Gaia likes it cold, and the state of glaciation is closer to the climate goal of Gaia than our present warmer interglacial. The reasons for thinking a glaciation more suitable for Gaia are these:

About 80 per cent of the past million years have been glacial, and during the glaciation the CO_2 in the air fell in abundance to less than 200 parts per million, about half its present level of more than 400 parts per million. Using evidence from the analysis of Antarctic ice cores, a team at Grenoble, in France, confirmed that global temperature and CO_2 abundance are closely connected. We should honour those pioneering French and Russian scientists who worked at the most remote of Antarctic sites drilling ice cores to gather the evidence that enabled us to look back in exquisite detail at the climate and atmospheric chemistry of Gaia over the past million years. These were, I think, among the most valuable scientific observations humanity has so far made, and more than any other give a glimpse of the physiology of our live planet. The fact that the Earth's temperature now cannot be calculated from their observations is no criticism whatever. We are the bad scientists, not those pioneering observers. We have erred by assuming wrongly that the polluted air we breathe now is no different from the clear skies of the last ice age, and we have erred in other ways in our naive assumption that the Earth now is the same as it was during the ice ages. In the interglacial period from about 12,000 years ago until 1700, the input of CO_2 to the air was more or less constant, and was determined by the escape of the gas from the Earth's interior by tectonic processes such as volcanoes and sea floor spreading. It is true that all living organisms, even vegetation at night, emit CO_2, but the net balance for a stable biosphere is zero emissions; the plants resorb the CO_2 when the sun shines. So why was the CO_2 in the air so low during an ice age? The most probable cause would have been the increase in mass of living matter as the Earth grew colder. At first this seems counterintuitive; we associate abundant life with tropical rain forests and know that the cold Arctic land is not a good place to grow crops. These are human perceptions and are probably false because we think only of the land surfaces where we live and ignore the fact that more than 70 per cent of the Earth's surface is ocean. Far from liking it warm, life in the ocean is now probably scarcer than in the ice ages. The physical properties of oceans are such that when the water temperature exceeds about 14°C the upper layer becomes separated from the colder waters beneath. When this happens nutrients present in the bulk ocean are unable to mix with the upper layer, and as a consequence it becomes an ocean desert with little or no life in it; this is why warm tropical waters are clear and blue – they are truly deserts. You may think of a desert as an area of the Earth's surface almost totally devoid of life. Such a state is found on the Antarctic ice cap, but only rarely elsewhere. The ocean and the land deserts in the tropics have visible vegetation, animals and fish but the density of life there is small compared with the forests or the soup-like waters of the Arctic Ocean and the nutrient-rich up-welling waters along the continental margins.

During the glaciations the ocean waters were cool and may have been rich with life, from microscopic algae to whales, distributed across most of the Earth. I made a mathematical model along with the American Earth scientist Lee Kump in 1994 that illustrates this hypo-thetical preference by Gaia for a cold climate, which I discussed and illustrated in *The Revenge of Gaia*.

In our blind belief that we now live in the best of all possible times, or at least that we did before global warming happened, and that glaciations are for polar bears and Eskimos, not for us, we ignore the fact that in the glaciation there was an area of land above sea level as large as, and in addition to, the continent of Africa now. Much of the land during the glacial periods was temperate. The freezing of all that ice, with glaciers down to London and the Alps in Europe and to St Louis in North America, to say nothing of the extension of Antarctic ice, removed vast volumes of water from the oceans and so led to a sea level 120 metres lower and consequently to the exposure of land that became available for occupation by land life. On this cool Earth with its richer oceans and more abundant land, there may well have been plenty of photosynthesizers (trees and algae) able to pump down CO_2 as they grew. Were it to happen now it would vex the North Americans, Chinese, Russians and Europeans to have another ice age, but the tropical nations might be greatly relieved. Whatever way we view it, it is far better to accept the natural response of Gaia – a move to a colder climate – than to accept the human blundering into an ever hotter and more barren Earth. The next glaciation after we had settled again might be a better world for us too. When the next interglacial approaches, perhaps our efforts should go towards keeping the Earth cold: working with, not against, Gaia.

As we gather more detailed and more accurate estimates of the long-term climate and compositional history of the Earth over tens of millions of years, we see that despite the ever-increasing output of solar heat, the Earth tends to grow colder. As long ago as 1992, the ocean scientist Michael Whitfield and I published a small paper in Nature with the provocative title 'The Life Span of the Biosphere'. In it we argued that if the goal of the Earth system was to keep its temperature close to glacial levels, the only way open to it in the face of the ineluctable increase of solar radiation was to progressively lower the abundance of CO_2. It was almost succeeding, and CO_2 levels 17,000 years ago reached 180 parts per million, possibly the lowest since life appeared on the Earth 3 billion years ago. The rise of CO_2 abundance to 280 p.p.m. in the interglacial before the Anthropocene had nothing to do with the presence of humans: it was the response of an over-stressed system to a rise in heat received from the Sun due to small cyclical changes in the Earth's orbit and inclination (the Milankovitch effect).

How confident can we be about the progressive increase of heat from the Sun? Compared with the Earth, the Sun is a comparatively simple system: its matter exists in a single state, plasma, composed of ionized gas; there are no complications such as those caused by the liquid, solid and gaseous states of matter on the Earth. There are almost no molecules stable enough to exist even on the coolest part of the Sun, where the surface is 6,000°C, and certainly nothing as intricate, protean and heat-sensitive as life. Models of the Sun with its single nuclear reaction source of heat, the fusion of hydrogen atoms, are more believable and robust than those of the Earth, and they project a mildly exponential increase of solar radiation incidents upon the Earth. The rate of increase of solar heat is such that in about 100 million years from now, if the Earth's temperature regulation depended on atmospheric CO_2 abundance alone, CO_2 would have to be close to zero to sustain a glacial climate. This is far too low for plant growth; consequently the Earth system, in its present form, is limited to a lifespan of less than 100 million years. Of course, a more heat-tolerant form of life may evolve, and the more probable lifespan could be as long as 500 million or even 1 billion years. The US scientist Jim Kasting and his colleagues in a 1993 paper argued sensibly for a lifespan of this order, based on much more heat-tolerant forms of life. It is worth mentioning here a curiosity of simple self-regulating systems. When the flow of heat from the sun is less than we would wish for, the temperature tends to settle at high levels; but when the Sun's heat is too high and the flux of heat too much, as it is now, the system settles at a lower than expected temperature. When the surface temperature is just right and regulation is not needed, then the temperature settles at the optimum. This is true of simple mechanical systems, such as an electrically heated smoothing iron, as well as, apparently, for the Earth. Critics of Gaia among Earth system scientists have argued that the unimpressive temperature regulation of our planet suggests that Gaia is either a poor regulator or does not exist. This argument ignores the fact that the range of regulation, from an ice age to an interglacial like now, covers about 6°C. This is almost the same range as is shown by ourselves, from the 42°C of a fever to the lower side of normal 36°C. Physiological systems are not constrained to regulate scrupulously at a constant temperature. They regulate themselves at what is best for the occasion, and for a warm-blooded animal the best has been determined by natural selection.

When contemplating the factors that affect global temperature it is easy to forget that there is much more to be considered than the abundance of CO_2. In the real world the principal greenhouse gas that keeps us warm is not CO_2 but water vapour. Unlike what we call a permanent gas, the abundance of water vapour varies with temperature. If it is hot, the air can hold more vapour and more water evaporates from the ocean; if cold more condenses out as mist, cloud or rain. Below 0°C water exists in the climatically different states of a snow-covered land or an ocean, a surface that reflects most of the sunlight back to space and is powerfully cooling; or as tiny ice crystals high in the troposphere, such as the contrails of jet airliners, which powerfully reflect infrared radiation coming from the surface and have a warming effect. All these different physical states of water affect in subtle ways the reflection or absorption of radiant energy from the Sun, which in turn affect the Earth's heat budget and so the temperature. And we must not forget that clouds do not always form when the air is cold enough to condense the water vapour it carries. Also needed are cloud condensation nuclei – tiny droplets or crystals onto which the water vapour can condense; without these, air saturated with water vapour condenses as huge rain droplets that fall from a clear blue sky. The effect of water vapour on global temperatures is exceedingly difficult to estimate from first principles, but in practice it closely tracks changes in CO_2. So intricate are the connections between the factors affecting climate change that we need to stay sceptical about the projections of climate models.

Clouds in the air can reflect some of the sunlight and be cooling. This is especially true of clouds near the ocean surface, called marine stratus clouds. Gaia is also in the business of adding a silver lining to marine stratus clouds. My friends Bob Charlson, Andi Andreae and Steve Warren published a paper in *Nature* in 1986 with me, suggesting that the gas dimethyl sulphide, which comes from algae in the ocean, is the main source of cloud nuclei in the unpolluted air over the ocean, and wondered if this was part of a Gaian feedback involved in the Earth's climate regulation. The Catalan scientist R. Simo and his colleagues have confirmed this idea for the atmosphere over the southern oceans, but they and others have failed to do so for the northern hemisphere; the failure is because we, through industrial activity, emit ten times more sulphur to the air than the algae do.

But there are clouds at other levels in the air, and cumulonimbus thunder clouds go all the way from the marine stratus level to the stratosphere. The net effect of these middle clouds on climate is still uncertain. Cirrus clouds that inhabit the upper layers of the air, the mare's tails that herald a change of weather, although similar to the marine stratus, warm the atmosphere rather than cooling it, through their back reflection of outgoing infrared radiation from the Earth's surface. As I mentioned earlier, important among these cirrus clouds are the contrails of jet aircraft. These straight lines of cloud drawn across the sky have a global warming effect.

Those then are a few of the factors affecting our and Gaia's choices. I hope that this account helps to show how far we have to go before we can make serious attempts to regulate the climate. I hope also it will make you empathize with the complicated burden carried by the few truly able climate scientists who have to lead the way in a world ruled by a 'confederacy of dunces'. This is no exaggeration: in most of the world's nations, fewer than 1 per cent of the politicians and administrators are scientifically literate. And as if this were not enough to hamper understanding, most of the language of science is fragmented into at least 100 different dialects. Scientists, even good ones, only rarely understand their colleagues of a different department. How can they possibly convey their wisdom to a politician? Far worse than this difficulty in understanding, humans inevitably, as part of their evolutionary destiny, spontaneously act collectively as gangs, lobbies, learned societies, political parties and cronies. These all have a consensus view that, even with respected groups such as the IPCC, can be far from accurate.

6

Can We Stop Climate Change?

Recently an interviewer told me, 'Your book *The Revenge of Gaia* was the scariest book I have read.' I was surprised; I had not meant to write a doom story, but intended a wake-up call, a warning that climate change was real and deadly and could be a threat to all humanity in only a few decades. I based my warning, as did many scientists, on the reports of the IPCC. The questioner then asked, 'Was your book over-the-top, and have you changed your mind? Do you now think that we can relax and continue with business as usual?'

'No, I do not,' I replied. 'I see the threat is just as real; all that has changed for me is that I have less confidence in the present model-based projections of the future climate. We are now less sure about when and how it will happen, but there is little doubt that there could be dire consequences for humanity's current way of life if carbon dioxide and population continue to increase.'

What has changed my attitude since the end of the twentieth century is a loss of innocence. Twenty years ago I thought that we understood the disease of global warming, and I was confident that we could cure it. Now I see global warming as one consequence of the evolution of the Anthropocene – as inevitable as population growth, economic instability and the emergence of artefacts beyond our capacity to control or understand.

Our natural tendency to believe in anecdotal evidence and reject science makes it extraordinarily difficult for humans to make the right choice now about climate change. The cause of our frustration was brought home to me by the distinguished science writer Oliver Morton. In a discussion with him on geoengineering, I had raised the possibility that, growing tired of the slowness and confusion of global bodies like the United Nations, business organizations whose operations were hampered by climate change might decide to act unilaterally by themselves. For example, it might be possible to install on cargo ships travelling the world a simple wireless-controlled aerosol generator that could produce clouds on demand wherever the ships sailed. A consortium of businessmen could then turn these climate changers on or off at will. At first I thought this might be a better way of tackling climate change than waiting the endless years for the United Nations or groups of individual nations to reach a sensible conclusion. After all, the climate might have changed irreversibly while we waited, and I think we should certainly look on the more or less complete absence of reaction to the Kyoto agreement in this light. Oliver Morton brought me to my senses at a meeting of the Geological Society in 2011 with this thought: suppose that the unilateral climate changes succeeded in cooling the planet sufficiently to sustain business as usual, but like all therapies, it was not without side effects. There could be severe droughts in powerful nations such as India and China, and they might have reason to suspect that the geoengineering was not working in their favour. They might then threaten to sink by military action the cargo ships performing the cooling unless they ceased immediately. The likelihood of arguments of this kind is what makes the United Nations deliberations so slow, and I am not surprised that in 2011 they decided to ban all unilateral attempts at geoengineering.

The ethics of geoengineering is of course the concern of those who make the decision to do it or not to do it. But in the present state of knowledge of the Earth system and its climate almost any act of geoengineering will be undertaken blindly. We are not now geoengineering a natural world that we happen to live on, but a world that we have already partially disabled, albeit unintentionally. We face a dilemma somewhat like that of a physician who knows that aerobic exercise is a good way to lessen the risk of heart disease but that it is hardly the treatment he would recommend to a patient disabled by congestive heart failure.

This kind of ethical dilemma is all too common, and is well illustrated by the problem of trying to level a table by using a saw to shorten the legs one at a time. Just imagine the consequences for the table if the problem was put to a committee and each member given a saw. That is what we are doing with regard to the Earth, made worse by the fact that the committee is made up of representatives of powerful nations, each with its own selfish national interest. The endless attempts by international bodies since Kyoto to find a sensible answer to the problems raised by climate change illustrate this human side of our dilemma. I find it depressing that bodies of scientists can talk of their consensus about the extent of climate change when they should all know that a consensus among differing opinions is no answer to a scientific question. True enough, there can be agreement among scientists about the best value of a measurement, when ten or so competent groups offer values differing only at their fifth significant place, but for climate scientists to talk of consensus about their projections of climate change when they vary from 2 to 6 degrees Celsius of warming tells either of naivety or of irresistible pressures from national or other vested interests.

Politicians and environmental scientists, especially in Europe, seem to have the mistaken idea that merely stabilizing the abundance of CO_2 in the air will stabilize the climate, and that reducing it back to preindustrial levels would rapidly bring back the climate of my childhood. This idea is based on the hopes of politicians and ideologues, not on scientific wisdom. More seriously, it is driven by the short-term aims of entrepreneurs who see vast fortunes to be gathered from the subsidies offered by government, especially in Europe, for renewable energy schemes. We must not forget the high probability that it is all but impossible for us to reduce the input of fossil-fuel combustion products to the air rapidly enough to change the climate favourably. Even if we reduced all emissions of carbon dioxide and other greenhouse gases to preindustrial levels, there would be no rapid return to the past climate and the less turbulent Earth we once knew. As Sir John Houghton warned in *Global Warming*, the climate system can change only at its own speed. We have to recognize that we are not in charge of the Earth, and almost certainly not yet able to geoengineer it to our own wishes. If I am right to think that Gaia is in a way alive, then we might compare our primitive ideas on geoengineering with fitting a horse with skates so that it could better proceed on ice. Perhaps it could in time.

I suspect that many climate scientists are aware of the near impossibility of their task. This is what makes me ask: is the pressure from their political paymasters so great that objective answers to global scientific problems are impossible? Could it be that they have been driven by their national governments to give answers to wholly unanswerable questions about the future state of the climate? They often fail because the real Earth system is only predictable to a limited extent. How could it be so easily predictable when the natural drivers of the system, especially those in the ocean, are huge and still mostly unknown? Even the contemporary forcing of the climate by air pollution, the fickle market and the burgeoning growth of population are each of them only poorly defined. Unpredictability is bad for climatologists, asset managers and economists, but not necessarily for the rest of us. We can take comfort from signs that the changes of climate, population and economics are lessening, which if confirmed might herald the approach of a new steady state.

We are spontaneously moving to live in cities, and not because politicians or ideologues have urged us to do so. Present-day politicians seem to assume that we merely have to 'decarbonize' the Earth's atmosphere by sustainable development and the use of renewable energy, and global warming will be under control. There is a widespread acceptance of this policy by the bourgeois elite, but it is based on a concept no more sensible than that of Lewis Carroll's Bellman on his way to catch the Snark. The phrase 'sustainable development' arose in 1987 when Gro Bruntland, prime minister of Norway, had to bring together businessmen who wanted development with ideologues who insisted on sustainability. She made a political judgement worthy of Solomon when she said to them, 'Why not have sustainable development?' Neither party asked what it meant, but they have happily tried to do it ever since. In no way do I mean to criticize Gro Bruntland for her introduction of the concept of sustainable development. At the time she suggested it there was, and to an extent there still is, a belief that economic growth was as essential for economic stability and well-being as is oxygen in the atmosphere; and if it could be sustained in a benign way, this seemed wholly desirable.

I recall once, when I was an apprentice industrial prophet in 1960s London, asking economists and senior managers of one of the major oil companies why we needed growth – 'For what,' I asked, 'is wrong with a steady-state economy?'

Their response could almost be described as one of shock-horror, as if I had uttered an obscenity. One of them – my friend, Sydney Epton – said, 'Jim, you must understand that without growth there would be utter chaos; the whole system of modern economics is dependent on growth.'

Taken literally, sustainable development simply means growth. No wonder the term is popular with economists who see the sustenance of growth as just what they need for economic stability. The flavour of the term changed when it became a shibboleth of green ideologues. Gro Brundtland's wise and moderate proposal was made an excuse for an ill-conceived industry for producing forms of 'renewable energy' that were inefficient and uneconomic but supported by huge and undemocratically acquired subsidies. There is nothing essentially bad about the idea of renewable energy, but when it is enforced by dogmatic ideologues wholly ignorant of both science and engineering, it is potentially both dangerous and ruinous. Would you be happy to live close to a nuclear power plant operated on theological principles? Have you ever wondered about the consequences of a free and unlimited supply of clean, renewable energy? It would certainly solve the CO_2 emission problem, but do you suppose for one moment that present-day humans would use it wisely? As the psychologist William James said, 'We can never have enough without wanting too much.'

I think that the most disreputable of all remedies for global warming is that of terraforming Mars or the Moon so that we have a refuge for escape if we make the Earth too uncomfortable to inhabit. Such an idea is wildly hubristic. Even at our worst, humans surely will not so devastate the Earth and the life it carries as to make us consider so grim an option. Be warned that those who propose the Martian retreat may be of an elite who imagine that if they could terraform Mars and make it habitable for themselves, they could ignore the billions left to starve on a dying Earth. But never confuse the bad idea of terraforming Mars with President Obama's recent proposal to send astronauts to Mars later in this century. This could be a worthwhile objective and sensibly follow President Kennedy's successful aim at the Moon. Had it not been for those brave astronauts half a century ago we would not have seen the Earth as a live planet nor would she through our eyes have seen herself: our voyages of exploration into space have greatly enlarged our understanding of the solar system, the Earth and ourselves. By comparison, the expenditure of $100 billion on the International Space Station has been almost useless, at least so far. In many ways any attempt to terraform and inhabit Mars could be a monster version of the Space Station.

Never forget that the average temperature of the Earth is calculated from all of its climates, ranging from the icy cold of the polar regions to the hottest deserts – from about -90°C to 60°C, and that the average is dominated by the 70 per cent of the Earth's surface that is ocean, with a surface temperature limited to 2–32°C. It is no comfort to someone living in Baghdad or Dallas when the temperature is 50°C to be told that the average temperature of the Earth is only 16°C.

If global warming is as serious as the IPCC modellers say, then for most of us their hottest projection, a 6°C rise in global temperature by the year 2100, might seem to make the Earth almost as uninhabitable as another planet. But would it? A hotter Earth would still retain our familiar, and essential for life, atmospheric composition, with oxygen at its present abundance; and there would still be ample water (although perhaps less freely available than now). This is a far, far better environment than Mars. Should we try geoengineering this altered world back to what we had until recently? Or should we adapt to the new world and leave Gaia to regulate the new wilderness? Most of the tropical regions of the present Earth are already 6°C above the global average and are well populated now. True, living in the equatorial regions now without air conditioning is unpleasant, but in a changing climate cities are much less vulnerable to external heat than are individuals. If most of us lived in cities, as it seems we soon will do, then the regulation of the climate of these cities might be a far easier, more economic and safer option in a hot climate than the regulation by geoengineering of the whole planet.

Climate scientists, philosophers, environmentalists, governments and the media all seem to assume that we have no option but to try to ameliorate planetary climate change, or join in with the rest of life and be roasted over the fire we have built for ourselves. Because we have so poor a record for dealing with global problems, such as climate change – and despite the drawbacks – modest geoengineering might be the most practical thing we could do to put off this fate. I still think it might be our only option if the onset of global warming goes much faster than predicted. Some geoengineering options, such as changing the Earth's surface reflectivity, could be relatively rapidly deployed, and if we made the low-lying ocean surface clouds more reflecting, it could be the basis of a more promising geoengineering proposal – one that could rapidly be shut off if unexpected hazards threatened. John Latham of NCAR put it forward as a counter to global warming. As with all inventions, the first embodiments rarely achieve a practical and economic optimum of efficiency. The talented Californian engineer Armand Neukermans has made substantial improvements in Latham's method for making an aerosol of sea water, and when fully reduced to practice it may offer the first economic and practical method of geoengineering.

We have to ask ourselves the painful question: are we seeking the survival of the largest number of humans, regardless of their condition, or seeking the survival of as many as we can keep in an acceptable condition? Should we aim to have a sufficient number survive on an Earth which we have tinkered with by geoengineering or by a reduced carbon diet? Would it not be a better strategy to have enough survivors in comfortable and civilized climate-proof cities?

When contemplating the future we must always remember the past. The Earth's climate 55 million years ago, in what geologists call the Palaeocene-Eocene Thermal Maximum (PETM), was about 6 to 8 degrees Celsius hotter than now. The north polar regions were tropical, and the fossils of crocodiles have been found in what was then the Arctic Ocean. I like to speculate that the climate system of the PETM resembled that of a pair of intact heat pipes, each one of them a single anticyclone (an area of high barometric pressure called a Hadley cell) going from the Equator to the North or South Pole. With such a system the Earth would be uniformly tropical from the Equator to the Poles. At present the Hadley cells are the anticyclone regions sustained by equatorial sunlight that are the source of the trade winds, once important for large sailing ships. These anticyclone cells go no further north than 30 degrees. Not surprisingly, if a wholly tropical

Earth happened in the PETM this might explain why there seems to have been no great extinction of species. Interestingly, from a purely physical viewpoint, the concept of a uniform-temperature planet, with a less complicated climate system, is confirmed by the present meteorology of Venus. We must also keep in mind that usually global warming, such as that thought to be happening now, does not simply make everywhere uniformly hotter. The increase of heat occurs mostly at the Poles and hardly at all at the Equator.

We may have wasted valuable time, energy and resources by trying to grapple with climate change on a global scale. It sounds good to try to save the planet, but in reality we are not thinking of saving Gaia, we are thinking of saving the Earth for us, or for our nation. The idea of 'saving the planet' is a foolish extravagance of romantic northern ideologues, and probably much beyond our ability. If I am right to think that our species is evolving maybe to become part of a planetary intelligence, then our most important task for Gaia, as well as for ourselves, is to ensure that enough of us survive to sustain our role as the first species to sense, think about and act to oppose adverse environmental change. We should be strengthening our defences and making a sustainable retreat rather than trying to 'save the planet'. We are not yet clever nor determined enough to serve in this way, but we could still be the progenitors of those that can.

The history of the human race reveals the possibility of population bottlenecks, when our numbers were reduced to only a few thousand breeding pairs. If this did happen then we are fortunate that our ancestors were brave and did not sink into a slough of despair. This is why I hope that we follow their example and keep in mind our most important task is to survive and pass on the genes we inherited.

Over the past forty years I have, although vicariously through the eyes of astronauts, seen our planet from space and seen how beautiful it is compared with its siblings Mars and Venus. If you think as I do that we are an organic part of the Earth, then perhaps our intelligence is a property of the Earth. This is why it is so important that we survive.

7

The Evolution of the City

Steven Pinker, in his recent book *The Better Angels of Our Nature*, says that 'the State's monopoly on the use of force in a territory reduces anarchy and thereby calms the "Inner Demons" that evolution has hardwired into us.' Pinker draws on extensive genetic, anthropological and paleontological evidence, and argues that the more organized the state is, the less violence there is. Could the present-day tendency to move to the cities, which become effectively states and sometimes, like Singapore, nation states, act as negative feedback on communal violence? And is this an evolutionary step, something that we have acquired and seem to share with the social insects?

I used to think that insects with their tiny brains could never think as we do, or be as aware of their environment. I saw them as hardwired automata, organic machines that did everything according to the inherited instructions of their genes. Recent investigations have shown that this idea might be wrong. As Lewis Thomas said of ants, in his book *The Youngest Profession* (1995), 'They do everything that we do except watch television.' I now wonder if social insects like bees and hornets, which might have good vision, would indeed watch TV programmes about their own world. Neural transmission in the bee brain may be 1,000 times faster than in ours because the distances involved are about 1,000 times less. That means the bee brain, a mere milligram in mass, may have the effective capacity of 1 gram of a brain like ours. Enough perhaps to enjoy a TV programme of hive news about a royal wedding. Just what the worker bees want! Most of them are females who share 75 per cent of their genes and have good reason to dream that any one of them might one day be promoted to be the Queen. This may seem whimsical, but having watched on several occasions the preparations for the nuptial flight of a queen ant and her bevy of competing males, I must admit to being carried away by the sense of occasion that attends these ceremonies.

The survival of the air-conditioned nests of termites in the Australian desert provides a fine example of how we might approach the problem of survival in a hotter world. A few years ago Sandy and I travelled on the Ghan, a train that crosses the continent of Australia from Adelaide to Darwin. The savannah and desert that we travelled through was populated everywhere with the skyscraper nests of termites. Was this how Gaia had geoengineered the Australian continent after adverse climate change? The invertebrate part of the desert ecosystem needed a cool and moist environment in which to live and reproduce, and they did this by retiring to the cities that they had built. As we will see, this may be an example we can follow. Whatever its form, the nest could be a lot more energy-efficient than the present-day heating and air-conditioning of individual houses. The most important question is not so much the source of energy, as using it efficiently. The nests could, for at least the first part of this century, use methane for heat and electricity, or nuclear energy. Later they might use solar thermal or tidal energy, when these two sources of energy are over their teething troubles. The main problem with energy is to acquire a reliable, economic and long-lived supply that has no adverse side effects. It is long past the time that we wound down the use of coal, oil, tar sands or indeed any high-carbon-content fuel. The use of these fuels to back up the huge inefficiency and intermittency of wind energy and other inefficient but profitable-through-subsidy energy sources, should probably also be banned.

The first cities we know about evolved about 5,000 years ago in what we now call the Middle East, in particular the area around and between the two rivers Tigris and Euphrates. This was a rich and productive agricultural region and is often called the Fertile Crescent. It seems likely that the climate in those times was different from now. We know more about the changes that happened in a similar region of south-west Pakistan, the Harappa. That then was a well-watered and productive region kept moist by the monsoon rains.

Later, possibly as a result of the destructive effect of poorly managed goat farming, the natural vegetation died along with the rainfall that previously had sustained it. The land then became the desert that it now is, and the farmed land contracted to an area that could be irrigated by water from the rivers. Some cities adapted; others were abandoned. A similar course of events may have happened in the once Fertile Crescent. Evidence suggests that similar cities appeared in other parts of the world, especially the Americas, and other cities may have existed on land now inundated by the 100 metre or more rise in sea level that accompanied the present interglacial.

The drive to live in cities is greater than would result from a mere reduction in the amount of fertile farmland. Anyone who has worked in a farming region, unless he is a farm owner, knows the attraction of city life. Work on the farm can be hard and monotonous, and if animals are kept, it occupies every day of the week. Life, even on a good farm, resembles that of a large family in a small house where naturally there is a strong urge among the elder children to escape, usually to the city. Never make the mistake of thinking that being streetwise and working in a city requires greater intelligence than work in agriculture. A good farmworker on a mixed farm, part arable and part livestock, has an amazing range of skills. They may exceed the mechanical skills of most city-based mechanics. The sheer diversity of machines, from milking machines to tractors and their numerous 'peripherals' such as ploughs, harrows, tedders, grass cutters and binders would baffle even graduates from today's universities. As if this were not enough, there is the need for empathy with farm animals. If, as a city dweller, you regard farmers who raise their livestock to be slaughtered for food as cruel, go to a farm at lambing time and see the care given to the mother sheep and newborn lambs. Would you go out in a blizzard at 3 a.m. to be midwife to a sheep? Knowing that live sheep and lambs represent money is a powerful motivation; but you will soon recognize, by his lack of success, the bad farmer who cares only for the money and not for the welfare of his animals. You may wonder how a farmworker with probably no more education than that provided in a village school can manage so well. I have often wondered about this, and concluded that the farmworker when a child undergoes an extended, but unscheduled and unofficial apprenticeship training in farming. The young growing up on farms around where I live seem to learn by doing. They look on the new tractor sitting in their own backyard as something quite marvellous and far better than any toy; they put enormous pressure on their fathers to let them drive it. Consequently most of them can drive a tractor and know all about its peripherals by about the age of ten. The Health and Safety Executive disapprove of this unofficial training, and to some extent it has been made illegal – justifiably, because accidents do happen to underage farmworkers more often than they should. Despite this, these ad hoc apprenticeships are a wonderful way of learning the craft of farming. Perhaps they could be combined with teaching in schools the fundamentals of farming and an appreciation of simple rules that would lessen the chance of dangerous accidents.

In modern society we may have made the error of being overprotective. I began my apprenticeship for science in a similar way to those young farmers: there was plenty of danger around but not of the kind that present-day children face. Few, if any, of my contemporaries died or were injured by drugs, knives or guns, but it was easier for me as a child to be electrocuted, or poisoned by lead from the sweet-tasting lead paint that was everywhere. The 240-volt electricity supply too often came along poorly insulated and perished rubber cables. I shall never forget the electric shock that threw me from my bath at twelve years old, caused by reaching out and touching a radio. It would never happen today, but in 1931 radios were manufactured with their metal chassis directly connected to the live side of the electricity supply, and touching any metal part, especially when well-grounded in the bath, was dangerous. I might have killed myself by this carelessness with the radio, but on the other hand the lesson I learnt from the shock, which was painful, taught me to regard with great respect the large and lethal transformers, capable of delivering 4,000 volts at several amperes, which were typical of the high-voltage electrical equipment I used during my working life. I had similar near-misses as a child with poisons. I recall chewing, out of curiosity, a small piece of leaf from the deadly nightshade plant, *Atropa belladonna*; it punished me with a day of misery from a dry mouth that felt as if full of sand and eyes with pupils fully enlarged, making vision unpleasant. These near-misses were worth more to me than hazard warnings or legal restrictions.

So why do we move from the full, rich life of the rural part of our world into the cities? And why do we willingly give up the fulfilment of mastering all those skills, and the delight of living in the country-side and sharing it with what is left of the natural world? It needs strong forcing, something powerful enough to override instinct and wisdom. I think it is mainly that powerful evolved trait, our very human desire for status. Do you remember from childhood the story about Dick Whittington, who when puzzled about whether to return to the countryside, chose to return to London and become its Lord Mayor?

So what do we find when we reach the city? We find a place with streets filled with shops and merchants, a place where skilled crafts-men learn and apply their trades, where tools are made, where disputes are settled and purchases of land are arranged. It is also the place where ideas can develop and philosophies evolve; where money and power resides; and the home of religions, of politics and – since we are tribal animals who like to have a leader to follow – of our ruler. This is most manifest with a civilization, a collection of cities united usually by a common language, and often a common ethnicity. The capital city houses also the palaces of kings, queens and presidents, the cathedrals of religions and the academies of the arts and sciences; and gluing it all together, the news media and the many houses of entertainment. The entire city is a place where upward mobility is possible on the status escalator.

For myself, a loner by disposition, the move to an urban existence is repellent, but then I felt the same way about having to go daily to school, a place I saw as a prison and an unjust punishment for the crime of being young. We adapt, and the worldwide move to an urban existence is by no means wholly bad. We evolved over the past million years from hunter-gatherers into farmers, and then small towns appeared, at first to serve and service the nearby farms. During the past 300 years evolutionary inflation has driven the growth of indus-try and has made the small towns into cities, which then accelerated their growth, took over much of the surrounding farmland, and like a magnet drew in the people of the countryside. Are we becoming a new species: social humans? Are our cities taking on the social structure of the nests of bees, ants and wasps, and becoming self-sustaining and self-regulating entities?

According the World Health Organization (WHO), 60 per cent of the world's population lived in an urban region by 2010, compared with 20 per cent in 1910. At the present growth rate the estimate for 2050 is 70 per cent. Most of us appear to be moving spontaneously to live and work in large cities, although currently most cities are sprawling settlements.

Apart from its mountainous areas, in Wales, Scotland and Northern Ireland, Britain is becoming one large city with a population of 63 million and growing. No wonder it is difficult to drive across – almost as difficult as New York or Los Angeles. But to make the whole of the United Kingdom into a single city does not seem to me to be a sensible option; I doubt that it would be possible even to make England a city. To cope with climate change, cities need a small footprint. Pre-1914 London, before the suburbs began to expand, was probably near the limit of size for a self-regulating city. The widely dispersed cities of the USA have population densities comparable with that of England as a nation – about 1,000 people per square mile – a density already recognized as too low for a city. The inflated area of US cities was made possible by affordable private transport, cheap fuel and inexpensive land, but it could not last indefinitely in the face of rising fuel prices. The unwise expansion of American cities like Los Angeles and Houston was also driven by the urge to escape the inner-city areas, which were either impossibly expensive or deeply deprived as a consequence of unwise political engineering. If climate change progresses as projected by the IPCC then American cities are likely to shrink in area and increase their population density. The same is true for much of the first world. For a world experiencing severe climate change and the social disruption it will bring, the UK is surely overpopulated. We did not have enough land to feed ourselves in the 1940s war. Now we have a larger population and less land. The answer may lay in high-rise, high-density, climate-controlled cities like Singapore. The English dream of a house and a garden is vanishing as fast as the empire that once sustained it.

*

I remember hearing in 1960, while still living in the cool climate of England, that the people of Houston, Texas, had succeeded in air-conditioning a whole street. I had spent time in Houston in the summer, and knew how good it would be to live in such a street. When I went there in 1961, I found that there was indeed an air-conditioned street, but it was what the locals called a 'mall' – in reality a consortium of air-conditioned shops occupying a single large building laid out as if it were a street. It worked, and clearly the costs were not so great as to prevent it being copied in cities across the United States. So if things do get bad with climate change, why not air-condition a whole city?

Would it not be easier for us to survive global warming in purpose-built cities rather than try to air-condition the whole planet either by geoengineering or by attempts at what is called sustainable development? If it should turn out easier, more economic and require less food to resist global warming by retreating to the nests, then the fact that people are moving spontaneously to live in cities should be seen as providing a wonderful opportunity. More than this: if we were wrong and global warming does not happen, the move to cities might be no great loss since we appear to be doing it anyway.

Is it really easier to climate-control cities than the whole Earth? I am not suggesting that funds are provided so that everyone living in, for example, Lagos has ample food, fresh water and air-conditioning. What I have in mind is climate control of a whole new city for the people of Lagos, which would then be like a city in a pleasant sub-tropical climate where personal air-conditioning is not needed. This subject has already been touched on in Lord Rogers's excellent book *Cities for a Small Planet* (1997). We have tended too much to treat the problem of climate change from the top down with ideas like carbon capture and reflection of sunlight. Perhaps instead we should consider the survival of people within climate-regulated cities, and the survival of the planetary ecosystem either as a single unified system or as two or more separate ecosystems. The Earth system has experienced long, hot spells in the past and repeatedly shown that it can sustain planetary self-regulation at a habitable state, although not necessarily a state comfortable for humans now.

The optimum temperature for the majority of organisms

is somewhere between 20°C and 30°C. The present global average temperature is at least 5°C less. As I explained in Chapter 4, Gaia likes it cold for cybernetic reasons, not because its constituent species voted that way. But because the Earth is about 5 degrees cooler than the optimum for growth we may have a huge reserve of cooling available to cope with global warming. The equatorial forest around Singapore, flourishing at 12.5 degrees hotter than the average, suggests that the cooling bank has hidden reserves.

Perhaps the world population should prepare to replace their inefficient sprawling cities with efficient compact cities designed to sustain an optimal internal climate, and leave the land and ocean to the Earth system to regulate as it has always done. I suspect that this would be much easier and far less costly than attempting by geoengineering to air-condition the whole planet at a temperature suitable for us but probably wrong for Gaia.

Adapting to greater heat by moving to air-conditioned cities and leaving the regulation of the climate outside the city to the Earth system is like wearing warm clothes in the Arctic. No one would be so foolish as to try to warm the outside Arctic air by a conflagration of bonfires, so why try to change the planet when we can live in the comfortable climate of a city?

Singapore is a large city (5.6 million people). It is prosperous and orderly with a high standard of living in an external climate twice as hot as the worst model prediction of global warming this century. It has a natural climate 12.5°C hotter than the global average but is kept cool enough for its inhabitants without their incurring onerous expense. It is vastly more expensive to air-condition a planet at a comfortable temperature than it is to air-condition a multitude of individual homes scattered across the planet; and far less expensive still, as Singapore has found, is the air-conditioning of the interiors of large high-rise buildings in a city.

Singapore, despite the heat and humidity, is one of the more desirable cities to live in. Not only this, but the natural environment around Singapore is not barren desert and life appears to be flourishing despite the torrid heat. It is true that since its modern development by Lee Kuan Yew, Singapore has been dictatorial not democratic, but then in the Second World War democracy was suspended in several previously democratic nations for the duration of the war. Perhaps a similar suspension of democracy will be needed when climate and other changes become as serious and as deadly as a major war.

So what would the new cities look like? I would imagine that whatever the first are like, the mistakes in their design would drive rapid evolution: evolutionary inflation is likely to make the corrections rapid. Surely cities built for survival would evolve in ways tailored to the local and regional climates and the availability of food and water. Constructions that were no more than enlarged and extended malls would compete with domes and underground cities. A factor in the shape of the city would also be its geographical location. We might go one stage further and make our habitations on a hot world simply well-insulated high-rise buildings in a more communal and better-planned city. Singapore and Hong Kong have already taken this step and are examples for thoughtful city planners throughout the world. Whatever the form of a city, personal transport would have to be a lot more efficient than now and entirely free of air pollution of any kind. The absence of weather in the city nest would allow the design of simpler and lighter electric transport. If possible, all essentials such as shopping centres, hospitals, schools, universities and airports should be within walking distance of homes – not more than two miles. Sandy and I find this distance quite practical in a suburb in America, despite my being well over ninety years old.

In a hot world our largest problem will be the supply of food and water. A great deal will depend on how we grow our food, and there are many options. We can go on as we are now with farms hacked from the natural forests and farmed sometimes rather inefficiently, or we can use intensive agriculture with the smallest possible footprint and leave a maximum of land for Gaia. We can even put our farms in nests; indeed, this is done extensively in the Netherlands. The horticultural greenhouse is a nested farm and it can be air-conditioned to a huge extent, even changing the gas composition of the air by adding CO_2. Beyond this, there is the biosynthesis of food, already shown to be practical although not yet widely acceptable and economic. Many of us even now eat biosynthetic food, made from the tissue culture of fungi and to be found in supermarkets labelled 'Quorn'. Such biosynthesis or greenhouse culture could be supplemented with tasty and nutritious items grown by horticulture, within or outside the nest.

Water may be less of a problem because of the huge volumes to be regained by preventing wastage, and of course the huge recycling potential of large cities. London was engineered this way as long ago as Victorian times, when a complete waste-water system was installed; amazingly, water coming from the cold water taps in almost every London dwelling has passed through the sewers of seven towns upstream on the River Thames. It is still safer to drink than the tap water of many modern cities. For example, in a Scandinavian city in 2007 we were disturbed to be informed that tap water was no longer drinkable because of contamination by the protozoan *Giardia*, one that causes an unpleasant intestinal disease. This was a single event and soon dealt with, but I cannot imagine this happening in London's massively recycled water.

As I mentioned earlier, Wilson proposed that for social insects the nest may now be wholly or partially the unit of selection in further evolution. This is a conclusion drawn from patient observations of the natural world. As we evolve a city, the basics of what we do and how we solve our problems is likely to have a certain amount in common with the natural evolution of the ants' nest.

If the idea of a city becoming like a living organism causes outrage in your mind, consider for a moment the evolution of the electrical or telephone network of a city. Wires extend from central nodes like the branches of a tree, each branch dividing into many smaller branches in a chorus of dendricity until the final tiny twigs reach into your home. How many of the workers who built these networks knew anything about the physics of electricity? I am not denigrating electricians; that greatest of inventors, the electrician Michael Faraday, also knew nothing of the electron, or why metal wires were needed to carry the electricity he had 'invented'. Indeed the discovery of the electron by J. J. Thomson came in 1897, long after the first use of electricity for lights, or even for trams. The city with its streets, wires and pipes grows dendritically, seemingly without a plan. Even more remarkably, each supply of water, electricity or gas is at a constant, closely regulated pressure or voltage, and the supply is sufficient. The investigation of major cities shows that the input of food by transport and the output of waste appear to self-regulate. The city, as Wilson and a few others of us have thought, is already a primitive super-organism.

Evolution in the context of Gaia brings in the material environment in a way that is absent from Darwinian evolution. Quite simply, if the evolution of an organism changes the material environment in a way that affects further evolution, then the two processes become tightly coupled. In the simple Daisyworld model the growth and spread of dark daisies, which warm themselves by absorbing sunlight, can warm the whole planet until it grows too hot for daisy growth; but if light-coloured daisies are also present, their growth and spread will cool the planet by reflecting sunlight. Competition between the two species of daisy leads in the model to a stable temperature and distribution of daisies.

So what will it be like twenty years from now? This must be one of the worst moments in history to be asked to predict the future. There are so many different forces operating simultaneously. We have just passed 7 billion people, economic growth is no longer a given, and there are looming unknown forces, such as the equivocal techno-logical changes forecast by Martin Rees. Much of what we do now every day in the world of telecommunication would only a few years ago have been science fiction. Did you ever expect to be able to record on your pocket telephone the sounds and vision of an escaped chee-tah walking down your street and send the movie live to your friend in Australia? This widely shared ability has suddenly made it far more difficult for evil regimes to commit atrocities; they would be seen live everywhere. There is no way of stopping this comparatively benign capability, nor the more dangerous things that computers make possible.

In the longer term, possibly global warming caused by our changes to the land surface and emissions of CO_2 will be as great as that of the PETM of 55 million years ago described at the end of the last chapter: but even so it need not be fatal to our great self-regulating Earth, any more than that period was. In other words, if we do noth-ing effective to offset global warming it will for Gaia be no more than a fever for one of us. I think it most unlikely that the number of humans surviving will be less than millions; this is more than enough needed for the survival of our species. (Our genetic history suggests, as I have mentioned, that there was a bottleneck when our total population fell to a few thousand, and that we all come from them.) I think also that if we move to cities efficient as termite nests, there will be a much better outcome. If Pinker is right, our present choice of city life could be a powerful, benign force moving us to a future existence in city super-organisms.

Singapore is only sixty-five miles from the Equator, and when it was first established as a port suffered one of the most intolerable climates on Earth. The average daily maximum temperature is 30°C and the humidity 75 per cent. As I related earlier, this is 12.5°C hotter than the global average now and about two times hotter than the most severe projections for global warming of the temperate zone. The lesson of Singapore shows that even if the climate worsens greatly and grows much hotter, a well-planned city offers life as good as or better than most cities now. I know that this may be far from what will actually happen. I wrote it to show that there are other options than the dismal choice of attempting to draw energy from impractical and damaging renewable sources. Whatever we are not, we are still an adventurous, creative and inventive species. Socrates's opinion that nothing interesting happens outside the city walls may at last come true.

8

Myths of Environmentalism, and Their Consequences

You will have been able to tell from the previous chapter that I think knowledge of the Earth's surface environment and a feeling for the natural environment should be part of our culture. I think it right that we hold it emotionally as well as rationally. As a scientist I accept that properly established facts must always constrain speculation, but they should never stifle our sense of wonder about the intricate and enthralling planet we inhabit. However, I grow increasingly concerned that our hard-won facts, established by proper scientific investigation, are being cast aside and replaced by the falsities of new dogma. We are retreating back to the world that Galileo endured, where any idea that did not conform to the religious beliefs of the period was treated as heresy. In what is now the world's leading nation, the USA, biblical dogma is too often accepted as fact in preference to biological science, and in much of the rest of the world truth depends on dogma of one kind or another.

It would be easy to think that we are evolving into a less intelligent form of human, but there is no evidence to support so dismal a view and the now well-established science of evolution denies the possibility. No, we are as clever or as ignorant as ever, but what has changed is that we are subjected daily, even hourly, to the enormous indoctrinating power of the media. If you doubt this statement, consider the success of television, social media networks, and even those otherwise benign search engines like Google. Almost all of them are paid by advertisers and opinion-formers. As Talleyrand memorably said, 'The truth is whatever is plausibly asserted and confidently maintained.' If

I tell you something three times you will believe it, said the Bellman, and usually we do.

The developed world today reflects the immense power of repetition by the media to rule our minds and the world. Marshall McLuhan, in his books *The Gutenberg Galaxy* (1962) and *The Medium Is the Message* (1967), repeated the warning. He even predicted the emergence of the Internet, but I doubt that in his lushest dreams he envisaged Twitter or Facebook. In this chapter I will try to show how through the flood of news from the media, our feeling for the world we live in has developed and then gone astray. Here is a simple example of how it happens:

Most of us are functionally innumerate and have no instinctive sense of the difference between a part per million and a part per trillion. If your glass of apple juice contains a milligram of potassium cyanide and you drink all of it, you will be unharmed. But if the juice contained a gram of cyanide you would most certainly die, and quickly. If you take a newly picked apple and happen to chew the pips you might swallow perhaps as much as a milligram of cyanide; it comes naturally from the apple and is harmless to you, but may well be a natural pesticide selected by countless generations of apple trees. Now consider a newspaper article that claims 'Scientists have found cyanide in apples' and then goes on to explain how poisonous it is and lists those unlucky enough to have died from swallowing cyanide. This is how we are misled, and why so many are fearful of all chemicals and all radiation.

Rachel Carson is often taken as the founder of modern green awareness through her powerfully moving book *Silent Spring*, published in 1962. She reminded the world that revolutions in industrial-scale science might be every bit as harmful as those of applied politics. After the massive disturbance of the 1940s war a major problem confronting the world was a shortage of food. Innocently, we thought that agricultural production could be substantially increased by developing antibiotics against the insect pests and parasitic fungi that greatly hindered the farmer's ability to supply us with food. Compounds like DDT (dichlorodiphenyl trichloroethane) already existed. DDT was used at the end of the Second World War to combat epidemics of typhus, malaria and other insect-borne diseases, and was impressively effective. Few bacterial antibiotics in use today are as free of side effects as DDT, but when it and similar chemicals invented to combat agricultural pests were used widely, and unwisely, on farms; and because nobody then knew the correct treatment regime, they caused profound local disturbances in biodiversity. The problem arose because of the amplifying effect of food chains. Small insects that had accumulated enough insecticide to partially disable, but not kill them, were an easier target for predators that might be larger insects, or small mammals like voles. These in turn would be prey for larger animals, and for birds like owls or hawks. Pesticides like DDT are only slowly metabolized and gradually accumulate in the animal's fat; because of this accumulation the pesticide concentration in the tissues increases at each step up the food chain. The owl that eats the vole therefore takes in a much larger proportion of the pesticide than did the insect at the start of the chain. The concentration of pesticides in the owl could soon exceed toxic levels, and the owl's progeny would be born poisoned.

This was the unintended consequence of pesticide use; it led the infant urban green movement to demand an immediate ban on the use of DDT and similar substances. A more sensible interim solution would have been to use lower doses, and at times choose not to interfere with the natural breeding cycles of birds, insects and other animals; and then start to develop more effective and ultimately less toxic insecticides. But young ideologues are impatient, and having read Rachel Carson's book they wanted an immediate ban on DDT. They, and urban green lobbies who claimed that all organic chemicals that contained chlorine were carcinogenic, made noisy demands until the US Congress obliged them and other nations soon followed. One consequence of this ban was a sharp rise in the human death rate from malaria and other insect-borne diseases in tropical regions.

It is as difficult to persuade a young green activist that chlorine-containing chemicals and radioactivity are natural and might even be good for us in small amounts, as it is to persuade a religious fundamentalist that sin is natural and that a little of it may do more good than harm. Fifty years ago I was beset by angry, buzzing green wasps who rejected my ECD because it contained radioactivity; and was at the same time in danger of punishment by a major oil company whose 'business as usual' was massively threatened by the harvest of measurements provided by the ECD.

It might seem that I was indeed in a serious conflict of interests by confirming the science that threatened to close part of Shell's business, the sale of pesticides and herbicides. But thanks to Victor Rothschild's humanity and excellence as a scientist (he was a Fellow of the Royal Society and a biologist, as well as the coordinator of science for the Shell Oil Company), he accepted Rachel Carson's findings, although he was angered by the emotive way she presented them. Shell responded by ceasing the manufacture of DDT and similar halogenated pesticides, but I recall Rothschild saying, 'She may be right, but no good will come of it.' He was shown to be prescient by the rapid rise in deaths from tropical disease that followed the ban on DDT. Interestingly, the first scientists to use the ECD to measure DDT in the environment were at the Sittingbourne laboratories of Shell, and they did so before the Carson bombshell burst.

So sensitive is the ECD that DDT was subsequently found in infinitesimal quantities in mother's milk in Finland and in the fat of Antarctic penguins. With the hugely respected Rachel Carson as the authority, journalists and TV producers soon had a global-scale scare story. It mattered nothing that the amounts found were far less than toxic. Finding any of it in human milk or in the remote life of Antarctica was strong meat also for the American eco-socialist Barry Commoner and the many others inclined towards varieties of primitive Marxism: they now had a strong new argument against capitalism. The lack of objectivity of these lobbies was confirmed by their persistent failure to condemn the massive pollution and mephitic industries of the Soviet empire. Neither Rachel Carson, nor the green movement – nor the US government – seemed aware of the dire human consequence of banning the manufacture of DDT and its lookalikes before substitutes were available. So fragmented is human knowledge among advisers, scientists and experts that no one told either Carson or Congress that banning it would lead to a rapid rise in the death toll from malaria and other tropical diseases. In 1963 malaria was about to become effectively controlled. The insecticide ban led to a rise in malaria deaths to 2 million yearly, plus over 100 million disabled by the disease. The World Health Organization (WHO) estimates that more than 200 million of us now catch malaria yearly. Carson was right about the death of birds, but wrong to give the impression that the entire world was being poisoned by pesticides. An unfortunate consequence of the academic division of science into disciplines is that if we think of a free electron at all, we think of it as in the province of physics not biology. In truth, electrons are everywhere and an important component of living cells. The free electron slowly moving in equilibrium with the molecules of our living cells is in fact the simplest of all chemicals, and is essential for the transfer of energy within the cell. The free electrons in the mitochondria may react with poisons as readily as do those inside the ECD, and the reaction leads to the formation of potentially destructive and even carcinogenic free radicals. The ECD revealed the presence of such compounds as the polychlorbiphenyls, dioxins, parathion, tetra methyl lead, and mere traces of the vapours of the explosives used by terrorists, and guaranteed the immediate media coverage.

Good scientists need to wonder why people are so easily scared by chemicals and radiation. True, the public associates them with that most feared of diseases, cancer; but only because they are exposed incessantly to fictional doom stories whose untruth is rarely challenged by those of us who know the truth. We need to explain the wisdom of Paracelsus, the sixteenth-century physician who stated, 'The poison is the dose.' We need to reinforce the message of Bruce Ames, the Californian physician who warned as long ago as 1983 that naturally occurring carcinogens made by living cells were in almost all food, and were more than one thousand times as abundant as carcinogens from chemical industry. We also need to stress that one of the most unpleasant cancers, malignant melanoma, is caused almost wholly by sunbathing or sunbed use, and not by having a nuclear power station as a neighbour.

In the 1960s and 1970s physical and chemical threats to the local environment seemed large, but we thought that eventually we could overcome them. Now, in the early twenty-first century, we are no longer sure that we can surmount the real environmental threats that confront us. Global warming could be devastating, and seems to loom like the wave of a tsunami, but although people and their leaders must be aware of its imminence, there is little inclination to do anything other than business as usual, or label their energy products, like diesel fuel, 'green' to make them more marketable. Market forces are distorted by subsidy to favour renewable energy and disfavour nuclear energy, the one source available now that would lessen the danger.

Modern intellectuals often seem proud of their ignorance of science. I suppose this is an inevitable consequence of our tribal natures. Michael Shermer, in his splendid book *The Believing Brain* (2012), has pointed out that selection favours our belief in anecdotes over our trust in science. In ancient times you were less likely to be cast out of the gene pool if you believed in myths or stories rather than what, in those times, went for knowledge. I would like to think that this explains the contradictory behaviour of the environmentalists, for many of them are my friends. Shermer is a leading sceptical scientist and there is much in his writing that complements this book.

There is mass misunderstanding of the consequences of 7 billion humans trying by industrial means to reach the standard of life enjoyed in New York, London, Paris, Berlin or Tokyo. It is not merely that Jesus's remark, 'The poor will always be with us' was an ineluctable economic truth. Now there are twenty times as many of us as were there 2,000 years ago, can the Earth support so many, and for how long? Let me remind you that the breath of 7 billion people, our pets and our livestock puts into the already overburdened atmosphere 7 billion tons of CO_2 a year. Heating and cooling our homes, powering our endless journeys by fuel-using transport and manufacturing all we need, bring our total emissions to 30 billion tons of CO_2. Do we really think that we can reduce our carbon footprint and at the same time make everyone as rich as we are? What would you say now to a wealthy Swede who asked, 'Don't you care about the poor?'

If some greens are as false and rootless as George Orwell found their middle-class socialist grandparents to be, who are the genuine greens? I would define them as those who think of themselves as an integral part of the natural world. Pre-Carson, the country village that I knew as a child was a seemly part of the landscape and enhanced its beauty. I have a fear that what is left of our countryside will be finally trashed by the dash for safe renewable energy. To generate one gigawatt of energy a year from wind turbines requires one thousand square miles of countryside, an area equal to that of Greater London or of Dartmoor. We use in England about 60 gigawatts of electricity a year. If our only supply was from wind we might need to cover the whole of England with wind farms. This is the green satanic change I fear. And worse, were we mad enough to do it and rich enough to afford it, we would still be emitting far too much CO_2 from the carbon fuel we would burn during the 75 per cent of the time the wind was not set fair for the turbines.

Professor Dieter Helm of Oxford University has recently proposed that the most practical solution for squaring the energy and pollution problem for a small nation like the UK is to make electricity by burning natural gas in a thermodynamically efficient energy cycle. This way we could turn about 50 per cent of the energy of the gas into electricity, compared with less than 40 per cent when burning coal. Burning gas releases half as much CO_2 as does burning coal for the same amount of energy produced, and if it were used to make electricity, and the power stations were inside the cities, the waste heat from the power plant could be used for space heating. Energy from burning gas might then begin to approach the harmlessness of nuclear energy. Gas may let us muddle through; its footprint is nowhere near so dirty as coal. Making electricity by burning gas is probably the best that we can do at the present juncture; when and if an efficient green source of energy appears we can use it. But because methane is more than twenty times as strong an absorber of infrared radiation emitted by the Earth, any advantage that it has over coal is lost if more than 2 per cent of the gas leaks into the atmosphere. I was glad to read that a study in the USA of methane escapes from fracking revealed that less than 0.5 per cent of the gas produced reached the atmosphere.

We in the UK could draw all the energy we need from nuclear sources, do it safely, quickly, economically and independently of imports. All that prevents us is irrational fear and a huge stack of bad laws passed by ill-informed politicians that increase the cost and delay the building of a new nuclear power station by ten years. Remember the world's first set of nuclear power stations were built and used in the United Kingdom in the 1940s and 1950s without any of the fuss that attends the building of even a single reactor now.

I do not think that humanity is yet capable of serving the Earth in a 'green' way. We do a fair job as omnivores fulfilling our role in natural selection, and we have every reason to be proud that we were able to show the Earth how splendid she looks from space. My hope is that we survive and evolve further to the point where we are as much a part of our living planet as our brains are of us.

TRUTHS ABOUT NUCLEAR ENERGY

Consider the incontrovertible fact that we inhabit a universe that is nuclear powered. All the stars draw energy from nuclear reactions; out there in the cosmos there is no such thing as renewable energy or sustainable development: the second law of thermodynamics forbids it. Our cosmos is running down, and has been since it started 13.8 billion years ago. A great deal of its energy is still potential, and some is stored as gravitational attraction: you can see it for yourself released as light when a meteor crosses the sky, and it is the source of the awe-stirring energy released as star after star is shredded and swallowed by the black holes at the centre of galaxies. The fusion flame of a star like the Sun is lit by the gravitational collapse of a huge mass of hydrogen until the pressure and temperature reaches the fusion ignition point. The star then burns steadily, and the heat generated raises the pressure, expands the star until a balance is struck. This soon becomes a wonderfully stable steady state, and stars shine for billions of years. The human ability to draw energy by burning carbonaceous material in the oxygen of the air is an ultra-microscopic part of the energy of the universe, a product of life's evolution; and so it is like us, something transient. Our inventiveness and scientific knowledge have shown us how to draw energy from the power-filled atomic nucleus, instead of from the mere chemistry of atomic electrons circling the nucleus. Unfortunately, tribal animals that we are, the discovery happened at a time of war. We could have used this stellar gift to provide the clean energy we now need, but instead we made bombs of fearful destructive power. If collective guilt is appropriate, then it should be reserved for the act of using nuclear energy for war so that it now haunts us with guilt, not the joy it should have brought.

It is true that there were good reasons to be deeply concerned about the warlike image too often conveyed by the US military and political establishment. It was close enough in reality to the gripping entertainment of the film *Dr Strangelove: or How I Learned to Stop Worrying and Love the Bomb* to be frightening. All political movements have their share of extremists and lunatics whether left, liberal or right-wing. The Campaign for Nuclear Disarmament (CND) was no exception.

By the 1990s the Cold War was over and the threat of nuclear war receded, but the CND did not disband. It regularized its relationship with the young and still politically innocent green movement, and quite naturally it switched from a campaign against nuclear bombs to one against nuclear electricity.

It is human to love celebrations and the fun of solidarity, and much of the success of CND came from the seamless way it took on the spirit of the old Labour Party marches. I strongly recommend the book on nuclear energy by W. J. Nuttall, *Nuclear Renaissance* (2005), where he succinctly explains the tribalism of the CND marchers and how difficult it would be to show them that their anti-nuclear arguments were little more than student diatribes against their old enemy big business. But I do understand the need when young to believe in and join a cause. I felt it when I was a student, and never forgot my sense of being left outside when I found it impractical as a teenager to join the Jarrow March in 1936.

Jarrow was a ship-building community on the edge of Newcastle in north-east England. At the start of the twentieth century it built a significant proportion of the world's shipping, but it was ruined by the 1930s depression. The unemployed of Jarrow made an orderly march to London led by their MP, Ellen Wilkinson. There was great sympathy for the plight of the marchers and I was filled with a wish to join them and be a part of their cause. As a 16-year-old schoolboy I could not do this. A few years later, as a student in Manchester in 1939, I read George Orwell's *Road to Wigan Pier*, with its vivid personal observation of the rampant poverty of northern England, and could see for myself the malnutrition that was everywhere in the poorer quarters of the town. This was indeed a cause to march for, but by then we were at war and united in a very different cause: how to survive as an intact nation an attack from the aggressive forces of Nazism.

How different was the cause in 1987 from that of the students in the 1930s: middle-class and comfortable, it was so easy to transfer their slogans and cries for the abolition of nuclear bombs to the abolition of nuclear energy, and the new name 'Chernobyl' gave them their rallying cry. The most amazing lies were told, still are told and widely believed. Even the BBC solemnly repeated that tens, if not hundreds, of thousands of Europeans would die as a result of the fallout from Chernobyl. As with poisoning by pesticides and other chemicals, there was wilful distrust and ignorance; and the failure to understand the wisdom of Paracelsus that 'the poison is the dose' applies to radiation just as much as it does to chemicals. Newspaper editors and their journalists seem to care only that they can attach a number, any number, to the quantity of radiation here, or poison there, wholly regardless of whether or not it is dangerous. So in 2011 the fearful of Los Angeles purchased the city's entire supply of potassium iodide, the remedy for radiation poisoning with iodine 131. They bought it when the news media filled their minds with the idea that they were in danger from the Japanese nuclear accident at Fukushima. So far as I'm aware, the quantity of radioactive iodine from Fukushima that reached Los Angeles was too low to measure against the natural background radiation of the city, and utterly harmless. What the Angelenos and the journalists did not know, so great was their ignorance, is that the potassium of the potassium iodide they swallowed was more radioactive than the iodine from Fukushima. All potassium is radioactive, and has been since our planet formed: about 30,000 atoms of it disintegrate every minute in your body. Moreover, the half-life of radioactive potassium is as long as that of the Earth, but that of radioactive iodine is no more than a week.

If a lie can be defined as a deliberate act of deception, then almost all nuclear scare stories are lies. Respectable media, including broadcasters, frequently fail to tell the truth about nuclear matters. They can only be correct if the relationship between the radiation dose received and the probability of death is literally true. For small doses there is no evidence for or against this hypothetical relationship. But let us assume it is true and that deaths from radiation are strictly proportional to the dose received. Where are the corpses, where are the graves of those unfortunate victims? It is now, in 2013, twenty-seven years since the Chernobyl accident, and despite at least three investigations by reputable physicians such as those of the UN agency UNSCEAR there has been no measurable increase in deaths across Eastern Europe. What is the answer to this conundrum? I think most probably the anti-nuclear lobbyists and their incompetent or dishonest advisers are ignoring a more relevant measure of harm from radiation, which is loss of lifespan. Those exposed in Eastern Europe to Chernobyl's fallout, may, if the linear no-threshold theory is correct, die a few days or a few hours sooner than their natural lifespan; this is an entirely imperceptible and immeasurable quantity. The anti-nuclear lobbies are skilled at propaganda and use the much more emotive and scary word 'death' rather than the much less disturbing 'loss of an hour's lifespan'. To put these terms in perspective, we now know that smoking cigarettes for thirty years or more reduces lifespan by, on average, eight years. Smoking is still legal, but we all seem to fear nuclear energy without cause.

Our ignorance is deep and incorrigible. In 1956 Queen Elizabeth II opened the Calder Hall nuclear power station in Cumbria and was, I suspect, proud to have done so, for it was the first in the world. At the time, nuclear was the greenest and safest of all energy sources and the only one that did nothing to add to global warming. Calder Hall is now closed and being decommissioned. Unlike the old Labour government of Clement Attlee in the 1940s that initiated the building of those first reactors, Tony Blair's New Labour government of the 1990s saw everything nuclear as malign; and he and his fellow CND marchers presided over the decommissioning of our only effective carbon-free source of electricity. One of the most unfortunate examples was when the Blair government forced the sale of the company Westinghouse, makers of nuclear power stations, to Japan, and set up the process for decommissioning the UK's fleet of nuclear power stations. This folly was highlighted in the unseasonably cold spring of 2013 when, for thirty-six hours, the country's reserves of gas fell below the amount we needed to produce electricity, and the price soared as purchases were made on the spot market.

In the 1970s we in the UK were drawing nearly 30 per cent of our electricity from nuclear energy. The cost of our electricity bills was low and no subsidies were needed. If we had proceeded in the direction we were moving then, by now we could be like France, drawing most of our electricity supplies from this safe and benign source and emitting far less CO_2. Whatever went wrong?

Fear of nuclear radiation trumped common sense. As prime minister, Blair then signed for the nation the Renewable Obligation, which requires energy companies to pay the subsidies for the building of almost useless wind turbines and solar voltaic cells. As a consequence, we now have ruinously expensive supplies of fuel and electricity in the UK. Most of us are unaware how great the differential is in the cost of energy between the UK and the USA. It costs three to ten times as much to heat a house in Britain as it does to heat a comparable house in America, which overall has a colder climate. This huge discrepancy, which affects everyone and all our industry in Britain, is part of the cost of believing in renewable energy as if it were a religious obligation.

The most obscene example of harm done by environmental activists is the torrent of lies about the 2011 Japanese earthquake and tsunami. It was one of the most severe earthquakes ever experienced anywhere in the last 1,000 years, and was followed within an hour by a 16 metre tsunami, a wave of water moving rapidly as if it were a solid wall, that destroyed all in its path. Most of the 26,000 who died were killed by the tsunami. The Fukushima nuclear power station was so well designed that it withstood without harm the earthquake and automatically shut itself down as it was intended to do. An hour later the tsunami wave hit the power station, and again it withstood as it was intended to do. Unfortunately the wave put out of action the emergency back-up diesel generators of electricity that were intended to power the pumping of cooling water to the now inactive reactors; the cooling was needed because when a nuclear reactor turns off, heat is still produced for several days by the radioactivity of the fuel, rather as the ashes of a fire remain hot for some time after the fire itself is extinguished. Unless this heat is removed by the cooling water, the reactor is in danger of heating to a temperature high enough to melt the fuel rods. In two of the Fukushima reactors melting of the fuel rods did take place, and a small quantity of radioactivity escaped into the surroundings. It will be a long and expensive operation to return the power station to production. But, almost never mentioned by the media, no one was killed by the Fukushima accident and no one has subsequently died as a result of it, unless we add in the suicides of those driven from their homes by an overzealous application of radiation safety rules.

The world reaction, largely a consequence of the media acting like a pack of hungry dogs scenting an enticing doom story, was panic. A humane world would have expressed deep compassion for the Japanese people, who had suffered one of the worst natural disasters in memory. Instead, Germany and Italy immediately shut down their own nuclear power stations. It was a wicked act and revealed us as a hopelessly ignorant species. It would only have been justified had there been unequivocal evidence that the Fukushima reactor itself had caused the earthquake and tsunami. Even the worst earthquake Europe has experienced was a thousand times less severe than that in Japan in 2011. Many otherwise intelligent people assume that the Richter scale of earthquake intensity is linear. It is not, and a Richter 9 quake is 1,000 times greater than a Richter 6. Each unit of the Richter scale of earthquake intensity is ten times greater than its predecessor. Had a Richter 9 earthquake occurred in Italy instead of the 5.9 one at L'Aquila in 2009 the historic stone buildings of Italian cities in the earthquake region would have been reduced to rubble and huge numbers might have died.

The lies about Fukushima still reverberate in the media two years later. Ignorance has triumphed again. (Although as I wrote those words, the news media in the United States released a statement that advice from a US environmental agency to the Japanese government after Fukushima, to make a 50-mile exclusion zone around the power station, was a wholly mistaken overreaction.)

THE DEPLETION OF
STRATOSPHERIC OZONE

Before global warming by greenhouse gases became a public concern, we were apprehensive about the threat of the destruction of the ozone layer, that tenuous filter that exists in the stratosphere and shields the Earth's surface from the harmful rays of solar UV. The thin air of the upper atmosphere might seem unimportant to our welfare here at the surface, but it is the place where ozone is found in greatest abundance, and this thin layer of ozone absorbs the fierce ultraviolet light from the Sun; it serves to protect us from much of the skin damage that would be caused by exposure to its rays. In 1973 it was claimed by two American scientists, Sherwood Rowland and Mario Molina, in an impressive article in the journal *Nature* that chlorine atoms released from the breakup of CFC molecules in the stratosphere could threaten the destruction of the ozone layer. The CFCs were then in widespread use as the propellant gases for spray cans of deodorants, pesticides and many other home uses. They were also used as refrigerant gases, and probably the greatest escape of CFCs came from car air-conditioning.

As I have explained, had the ECD not been invented, a device that could detect and measure infinitesimal traces of the CFCs, it is doubtful that they and other dangerous compounds could have been found as soon as they were and before the publication of the Molina–Rowland paper. There was no proposal anywhere in the world to fund measurements of the CFCs in the air, and it was left to me to do it, funding it mostly from my own income. From my voyage on the ocean research ship *Shackleton* in the early 1970s, I discovered that the CFCs were present in the atmosphere of the whole world, and that they were accumulating apparently without any natural process for their removal. This was potentially dangerous because the CFCs are potent greenhouse gases, and although in 1970s their presence was low enough to make them harmless, I calculated that at their projected rates of emission, by the end of the century (2000), they could be a serious contributor to global warming. When Mario Molina and Sherry Rowland published their seminal paper on the probable destruction of the CFCs in the upper atmosphere by ultra-violet from the Sun, and the consequent release of chlorine atoms, the scientific community realized that here was a credible and potentially large environmental risk. We already knew that chlorine atoms could react catalytically with ozone, and remove it from the stratosphere; the CFCs were therefore potentially an immediate danger.

It was not long before activists were making strident calls for the immediate banning of CFC emissions, but as is often the way with environmental arguments, the 'ozone war' between believers and deniers began before much of the crucial evidence was known. As a scientist I strongly disliked the gut responses of both the believers and the deniers. I needed more facts to answer some of the questions raised by the Molina–Rowland hypothesis of ozone depletion – such as, for example, were the CFCs destroyed in the stratosphere by solar UV? The UK Ministry of Defence had been a customer of mine since 1965 and I had many friends in its security services. When visiting them on a wholly different matter I mentioned my difficulties in finding someone who would fly me into the stratosphere to measure the CFCs there. One of my MOD colleagues mentioned that a Hercules (C30) aircraft was due for a test flight to check its performance and suggested that I ask the pilot if he would mind giving me a ride when the test was being done, which happened to be in about ten days' time. The pilot and crew seemed to welcome the chance of doing some science as well as routine tests, and so I travelled with them on a short flight from Lyneham military airfield into the strato-sphere and back. Their aircraft was lightly loaded for the flight, and could reach the aircraft's ceiling of 45,000 feet. During the flight I measured the atmospheric abundance of CFC11 and carbon tetra-chloride as the aircraft climbed; as soon as we were above the tropopause at 28,000 feet, the level that marks the entry to the strato-sphere, the abundance of CFC11 and carbon tetrachloride began to fall, and by the time we reached the aircraft's ceiling it had fallen to a few per cent of the tropospheric abundance. My measurements con-firmed the Molina and Rowland hypothesis of a sink for CFCs in the stratosphere. The most probable explanation was that CFCs were indeed photolysing in the solar ultraviolet radiation and releasing chlorine atoms that could react with ozone. The findings were published in *Nature*, but interestingly neither the scientists among the believers, nor the believers in doom from the CFCs, approached me or asked about the paper. This was odd because normally the first thing scientists do when there is strong evidence from measurements supporting their hypothesis is to confirm the details and the quality of the observations; then, if satisfied, use them as evidence to support their hypothesis.

My guess, and it is no more than a guess, is that their failure to approach me was a consequence of science since the Second World War becoming a career rather than, as previously, a vocation. In the USA, which had become the world centre for science, it was more like a team sport played for prizes and recognition, not out of curiosity as in old-fashioned Europe. The reason my *Nature* paper was not cited or used as supporting evidence was, I guess, because I did not belong to a recognized team; worse, I did not belong to any team at all!

There is an entertaining account of the CFC issue in *The Ozone War* (1978) by Lydia Dotto and Harold Schiff, My personal motivation for seeking the presence of CFCs in the global environment was the pleasure of a long sea voyage combined with the wish to discover if they could be used as a tracer of human pollution – a way of discovering the size of our polluting footprint on the Earth. The outcome of the voyage was published in *Nature* in 1973, and in it I claimed that one fluorocarbon was accumulating in the atmosphere with no signs of any process for its removal. I found that the global average abundance of CFC11 was 50 parts per trillion, about 1 billion times less than the least amount considered toxic. Unwisely, I said in the paper that the presence of these compounds represented no conceivable human hazard. I did so to counter claims by Barry Commoner, that organic chlorine compounds were unnatural and toxic when breathed at any abundance. Although I was right to challenge Commoner's assertion, it was an unfortunate blunder to have done so only months before Rowland and Molina published their famous paper predicting that the CFCs were probably a threat to stratospheric ozone.

There was a risk, rarely mentioned by believers during the ozone war, of a pandemic of intestinal disease resulting from over-hasty action to ban CFCs. An instant ban, which the activists sought, and had achieved with DDT, could have soon rendered food storage difficult or impossible. It is easy to stop producing CFCs, but to find replacement refrigerants that are environmentally harmless takes time, and on a global scale this could be as long as ten years. I was glad and grateful to be asked to appear as a witness before a US Government Congressional committee dealing with the practical issues of the ozone war. I gave my testimony at the same sessions as did Sherry Rowland, and stressed before the congressional delegates that we still lacked a great deal of evidence on this atmospheric problem, offering as an example my own recent discovery that there was a natural chlorine compound, methyl chloride, in the air in greater abundance at that time than the CFCs. Neither the believers nor the deniers had known this, and could not take it into account; the outcome of the hearing was to give the CFC industry more time to develop safe alternatives for refrigeration. This has been done, and modern refrigeration is no longer a threat to the ozone layer, although we are still anxious about the huge infrared absorption of some refrigerants.

Some of the warriors fighting the ozone war were moved more by testosterone than by common sense, and were indignant at my success in persuading the American government to deny them their triumph of an immediate ban. It was not long after the congressional hearing before I found myself pilloried in the media as a bought man of the chemical industry and worse, publicly denounced at a Nobel symposium before the King and parliament of Sweden by the aforementioned Commoner. I was lucky that this happened before blogging became the medium for scandalous attack.

It must have been hard for academic scientists to believe that I had not been corrupted, because I was frequently seen in the company of Ray McCarthy, a senior scientist at DuPont, the principal manufacturer of CFCs. Ray became a close friend and provided me with data on the emissions of CFCs by the other chemical companies of the world. He died recently without recognition by the world community for his strength of character and the clear-headed strategy that enabled him to lead the industry to accept the Montreal Protocol banning CFC emissions.

In the 1970s the quality of much of the observational and theoretical atmospheric science during the ozone depletion affair was poor. Much more than now, there was an inexcusable faith in the predictions of half-baked mathematical models. The science was so bad that when orbiting satellites first measured and reported low values for ozone in the south polar stratosphere the information was rejected by the scientist 'believers'. They then programmed the data collection from the circumpolar satellites to ignore low values of ozone above the South Pole. They were so in love with their models and so sure that they were accurate that they considered the ozone hole over Antarctica impossible (see *The Hole in the Sky*, John Gribbin, 1988). The same thing happened with observations of CFCs at the surface: only those that agreed with the model could be accepted. Those who have faith in models or received ideas, rather than trust in well-made environmental measurements, are no longer objective scientists: they are no more than believers or deniers.

CONCLUSIONS

I was deeply involved personally in the debate about the danger posed by the CFCs to stratospheric ozone; so deeply that I find it difficult to write about it objectively. I surely hope that a historian of science will take on the task and start by wondering what that disciplinarian of science Karl Popper would have thought of the politicized scientists who adjusted the data reported by satellites to make them agree with their mathematical model projections. Although little has been said about the shameful conduct of science in the ozone war, we have learnt from it and the quality of climate modelling and observations now under way is far better. But have we yet, as a society, learned to use our intelligence to overcome the tribal and political prejudices that lead to bad science?

9

Human Nature and Tribalism

Assume that the climate scientists are right, and that during this century we may experience global warming as severe as 5°C. It may be that we can cope with this hot climate simply by retiring to insulated and climate-controlled cities. If we do this, and leave the land outside the cities to be managed by the Earth system or Gaia, then we may find that we can cope with the hot climates projected by the IPCC. Among the things we need to know is how the separated human cities will behave tribally.

Have we as animals changed much from our hunting and gathering ancestors? The anthropologists, who have examined groups of humans living in environments as diverse as the Arctic, the inner city and the tropical forests of New Guinea do not seem to think so. I remember a gathering at Raleigh-Durham, North Carolina, in 1975. Those attending included some who went on to be leading figures in the science and policy-making about global warming. It came about when Margaret Mead, the renowned anthropologist, asked William Kellogg, from NCAR, and me to help her organize a conference on 'The Atmosphere: Endangered and Endangering'. As might be expected from the eclecticism of the conference title, our discussions were not solely about the arcane physical chemistry of the upper atmosphere and the threat to all life on Earth from the erosion of that fragile shield, the ozone layer. During a coffee break at the meeting itself, she told me of a young man who lived a typical western suburban life in Port Moresby, New Guinea. He came from a hunter-gatherer family whose way of life had not changed in thousands of years; she had held him as a baby when she was a student anthropologist and lived in the jungle village with his family. In a mere twenty or so years, he had 'evolved' from his ancestral world to the suburbs of a twentieth-century city, and commuted to work daily by car. Evolutionary inflation has massively changed our environment and the way we live, but as individual animals we have changed only slightly in the course of 10,000 years. We are socially, it seems, now in the midst of a change as great as that of the evolution of solitary insects into the amazing nests of bees, ants, hornets and termites.

What most impressed me about the life of social insects was the suggestion that the disciplined behaviour of insects in their nests was sustained by status, just as we use a status ladder to train soldiers or young physicians and surgeons. Newly hatched bees appear to be given menial tasks such as clinging to the nest's entrance and fanning their wings to keep air flowing through the nest for ventilation, cooling, and perhaps keeping sufficient oxygen available. After a while they are rewarded by a rise in status to that of a carer for the larvae. Then finally they reach the high status of a skilled worker, allowed to fly from the nest and forage for honey and bee bread. Of course above them all is the Queen, in some ways like a constitutional monarch with ultimate status but little power. How is order kept in the nest? I had previously been taught that insect behaviour was almost entirely hardwired so that insects were almost indistinguishable from artificial automata, but recent research suggests although some hardwiring occurs, behaviour is also regulated, as it is in mammals, by gifts for good behaviour and corporal punishment for transgression: the higher-status insects will nip or pinch with their jaws those who break the rules.

For humans social status is everything. The English are often accused of having developed an elaborate and divisive class system, but I doubt that it was more divisive than the Hindu caste system or the way that degrees of celebrity determine status in the world today. Could it be that class divisions are evolved to sustain stability? Where the British differ from other Europeans is in enjoying periods as long as a thousand years without disruptive regime change; long enough perhaps for class divisions to evolve with stability as an objective?

I was raised from my birth in 1919 until I was six years old by my grandparents in Letchworth Garden City thirty miles north of

London. They were clearly and unequivocally working class, and would have been recognized as such by their characteristic north London way of speech. My grandfather had reached the elevated status of a skilled worker; he was a bookbinder who knew his craft well. He and my grandmother were proud that he had bound one of the sacred books exhibited in Winchester Cathedral. Although left-wing, my grandparents were happy in their place. I have always found it strange that the academic left, many of whom are from the working class, seem unaware that the status ladder does not come to ground at the working-class level. Many levels existed below and above that of my grandfather, and with these subclass distinctions went a great amount of snobbery. My grandmother, although intelligent and staunchly socialist, would refer to those in the lower working class as 'common', and despite living in the allegedly classless society of today, most of us still know what she meant. She died in 1941 and her views, which were widely held, indicate how recent is the present-day acceptance of the concept of egalitarianism. Huge changes took place in the western world during the twentieth century as the Christian religion slowly gave way to a secular belief in what Keynes defined as equality of outcome. Taken literally it seems to rule out any kind of class system, and is inconsistent with order in a city or an insect nest. The 'equal opportunity' form of egalitarianism seems less difficult, but both forms smell of demagoguery. Its inconsistency with order is nicely summarized by the apocryphal military adage, 'We may be equal in the sight of God, but not in this army.' I don't think that egalitarianism would ever work in a city super-organism. For many species of social insects, their urban nests are effectively army camps. Thus the *Eciton burchelli* ants are even called army ants, and spend their lives in bivouacs, or in never-ending campaigns. In several ways the insect nest is as good or better a model of a human city as any of our present-day computer models.

I can't help thinking that a large brain is not necessarily needed for the everyday exchanges of life, so why shouldn't a mouse feel just as strongly as we do? I take it as a working hypothesis that most mammals and birds have emotions similar to ours; sometimes with greater and sometimes with less sensitivity. This is not sentimentality on my part but the recognition that their nervous systems are close in structure and composition to ours, and might even handle neuronic software not so different from ours.

Although it is unwise to compare animal and human mental capacity with that of computers, the analogy may sometimes be useful. In 1981 I was the proud possessor of an HP desktop computer, a 9836C. I used it to write my second book, *The Ages of Gaia* (1988). It had a splendid word-processing program with a spell, grammar and style checker; best of all, the entire program was written in engineering basic and I could add or subtract lines of code when it seemed necessary. I mention this because the whole memory of the computer, operating system and programs occupied a mere 1 MB; the PC sitting before me now has 10,000 times more capacity and an operating system 1,000 times faster. Curiously, there seems to have been only a small improvement in word processing accompanying this 10-millionfold increase in capacity, so maybe that honey bee with a brain 3 million times smaller than mine or yours is much more thoughtful than we imagined.

Status is also expressed through our membership of groups, and the most important of these in evolutionary terms is our tribe. What is the significance of tribal nationalism in an evolutionary biological sense? When one of our distinguished evolutionary biologists, John Maynard Smith, was a guest at our Devon home in the 1990s, I asked him what he thought about the possibility of a tribal gene, and whether we could remove it if it did exist, and he replied that tribalism would almost certainly be distributed over a wide range of genes and it would probably be impossible to eliminate tribal behaviour by genetic surgery. Somewhat naively, I had been wondering if we would be better off without tribal genes – my Quaker upbringing had led me to assume that the world would be a better place and free of war were there no tribal forces to ignite conflict. We are most of us subject to fits of liberal naivety, and it was my turn to imagine that the genetic forces that led to tribalism were both easily understood and could be manipulated.

A glimpse at the intricacy of the evolution of tribalism was described in an article in *Science* in March 2011 by the anthropologist Kim Hill and his colleagues at Arizona State University in Tempe. The paper concerned an analysis of genealogical marital data, which shows that among the modern hunter-gatherers, monogamous sexual unions between men and women from neighbouring groups create networks of in-laws that spawn widespread cooperation and cultural learning. Hill and his colleagues wrote: 'Alliances between foraging groups were facilitated because the unrelated males all associate with the same female, who may be their daughter, sister, wife, mother or daughter-in-law. By friendly association with her, males begin to associate with each other.' Hill went on to describe a social system which encourages collaboration among genetically unrelated individuals, and suggested that it originated about 2 million years ago, as human ancestors began to hunt and gather foods that youngsters could not obtain for themselves. In this situation, females would have had an incentive to seek mates willing to stick around and provide food for offspring. Joseph Henrich, an anthropologist at the University of British Columbia in Vancouver, said that 'differences in social structure, not necessarily cognitive advances, allowed our species to cross the barrier to cumulative cultural evolution'.

Previously it was thought that modern foraging communities, and by inference prehistoric groups, consisted of many male relatives, with the women migrating into groups as marriage partners. Hill's analysis of the range of hunter-gatherer populations shatters that assumption. Using bands of modern hunter-gatherers, Hill and colleagues analysed data on more than 5,000 individuals from thirty-two different societies worldwide. Each society consisted of two or more bands of people who live together while moving about the landscape. Bands range in size from five to sixty-four individuals. The researchers concluded that three social features characterize hunter-gatherer societies, and make them unique to humans. First, both men and women are as likely to stay in the bands they were born in as move to new bands to find marriage partners. Second, adult brothers and sisters frequently reside together, along with many in-laws. Third, a majority of band members are genetically unrelated.

Bernard Chapais of the University of Montreal argues that this monogamy-based social structure encourages males to circulate freely among bands in which they have relatives and in-laws. He proposes that cultural innovations and traditions are thereby spread rapidly and unite the bands into larger tribal units. From such origins may have grown our current deeply tribal civilization. Tribal differences reveal themselves in almost all aspects of human intercourse, not just racial differences and tensions between nation states, but also tribal disputes over infinitesimal differences, such as those between team players of a sport, or between those who interpret slightly differently the dogma of the same branch of their monotheistic religion. It even happens in science, and nowhere more prominently than between the believers and deniers of climate change.

The believers in climate change are deeply tribal and often belong to the left-wing tribe as well, whereas the deniers, also tribal, tend to be right-wing. Some idea of the complexity of the interlinks involved in tribalism is given by the ease with which a football team, no matter how varied its racial content, can become tribal and be followed by supporters who indulge in football violence. Another puzzle is the occasional absence of tribal feeling among the combatants on a battlefield, whether it is on the ground, at sea or in the air. In wartime, numerous accounts exist of fierce but chivalrous behaviour between combatants, especially in desert warfare and in the aerial battles between fighter pilots. Of course, in these examples the battleground was not on home territory; from the combatants' viewpoint, the desert, the sea and the air were unoccupied territory. More uninhibited violence took place where homes and families were invaded.

I'm trying to argue here that while I was naive to assume that there is anything as simple as a tribal gene, tribalism is so strongly a part of our natures that there is little chance that a benign education, selective breeding or genetic manipulation can be used to alter it. Like it or not, when confronted with an invasive threat to our territory we will respond tribally. On the other hand, when confronted by a huge natural disaster, the idea that we all belong to humanity as if it was our tribe evokes little response, unless the disaster affects us directly. Of course I am talking here about individual personal responses to distant disasters. It used to be said before travel and instant communication became the norm that the death of a pet dog would move us more than the deaths of millions living in a faraway place. To some extent this has changed: the response to the tsunami in the Indian Ocean in 2004 was certainly great, and to an extent effective. But before we assume that we are growing more compassionate, we have to ask why there was almost no compassion for the Japanese people who suffered greatly in the tsunami in 2011. The answer is clear: the information we received for both these recent tsunamis came from the media. For the Indian Ocean tsunami a clear account of devastation and suffering was the news; for the Pacific Ocean event the media concentrated on the damage to the Fukushima nuclear power station, and after the first day almost ignored the loss of 26,000 lives and massive suffering as the tsunami wave devastated the coastal plain. We have not changed our capacity for compassion, but we can only respond to the news that is given.

It seems probable that the largest human group we can respond to without priming is our tribe or nation. The concepts of humanity or the human race sound good in political exhortations and sermons but are essentially beyond the perception of most of us. In fiction it has often been said that an attack by an alien force from outside the Earth would be needed to cause a united response. But the idea has happily never been tested.

That most elegant of models, the nest of social insects, clearly displays these tribal responses: stir a wasps' or hornets' nest to see the violent coordinated response; but watch them later repair the breach in the integrity of their nest with calm collaboration. Rarely are these different responses shared with nests in the same locality. Just as well for us: just imagine the terror of finding that all the wasp nests of your locality had formed a united wasp union so that all of them joined in the chase for you when you accidentally disturbed a single nest.

The most important problem confronting humanity is survival. We have to survive because we might now be one of the most important species on the planet, perhaps the only species that can directly affect Gaia's chances of survival. Just imagine that astronomers had observed a large planetesimal on course for a direct hit on the Earth; no other species but humans would have any possibility of seeing or deflecting it. At present we could perhaps deflect a small piece of rock on course for the Earth, but it would require a great deal of organization and a space-guard service to make even this minimum response possible. I like to think that we will reach the stage where, given the information about a possible planetesimal strike, most nations would pull together in the effort to deflect it.

We are important not because of what we are now but because of our potential as an intelligent, communicating animal that might eventually evolve as an integral part of the Earth system. But will we evolve to the stage where our tribalism and its offspring, nationalism, defer to planetary needs?

I find it helpful from an imaginary Gaian viewpoint to compare humans and other intelligent animals with the first photosynthesizers. They learnt how to gather energy from sunlight and were the progenitors of the plants and the intricate system that evolved around them and was needed to run the Earth system; their evolution led eventually to the evolution of vegetation in all its forms, from micro-organisms like algae to giant trees. The material substance of this vegetation could be regarded as biofuel, and it forms one half of a battery; the other half being the oxygen that plants emit to the atmosphere. Animals in all their forms, from amoebae to lions, evolved simultaneously to limit the chemical disequilibrium of the Earth at a level Gaia could handle. Just how well the balance is kept is measured by the fact that a mere 1 per cent increase in oxygen abundance would greatly increase the risk of forest fires. Interestingly, the procedure used to sustain the balance is the burial of 0.1 per cent of the carbon photosynthesized. This of course is fossil fuel. It was not buried so that we could benefit, it is a part of Gaia's self-regulation. This process has empowered the Earth system for much of the last 700 million years.

The animal–plant system is important through its capacity for energy transfer; our role has been to start the process by which Gaia awakens from her long 3 billion years of infancy and becomes aware. We give the Earth the chance to become an intelligent planet more capable of self-regulation, and with potentially a longer useful life-span. The human species may be the start of a comparably important information-transfer system, one that will powerfully and directly aid Gaia as she continues self-regulation and continues to export entropy to space.

It would be easy to assume that the future of the Earth and life upon it depends on the evolution of better cooling systems than we have now, as well as a more efficient adaptation to increasing heat. Some adaptation is already taking place, but the living cells of nearly all organisms can tolerate only a narrow range of temperature. The capacity of adaptation to solve the heat problem is thwarted by the fact that the heat from the Sun increases ever faster, remorselessly; mere adaptation is like relying on a loan shark to sustain your bank balance.

Natural selection is a slow process, but its inventive capacity has so far coped with the exponential increase of heat from the Sun. As the pace of solar heating quickens and our own excessive numbers and pollutions vastly increase the retention of the Sun's heat, it might seem that our future is grim. But as we have greatly increased our population and our consumption of fossil fuels and other resources, so we have also enlarged our capacity to invent.

As the bell of Nemesis tolls the end of the current Earth system, in the distance we hear the sound of trumpets. Coming to the rescue is an army of super-nerds. Slow-moving natural selection has at last evolved a species, us, which can invent and so increase the rate of evolution a million times. Now we have the luxury of time to choose between a move to an entirely new form of electronic life, or adaptation; or both, combined as endosymbiosis. We can now sense the dangers ahead and sensibly devise a means to keep the planet cool or adapt to increasing warmth.

Can we now save ourselves? As far as metabolism is concerned we are still fully integrated with the Earth system and likely to remain so for long enough to cope with change, but our intelligence is still at an infant stage and not yet capable of coping with its own mistakes. In particular, our inventiveness is not channelled, as it would be in times of war, but wasted on trivia and indeed anything that sustains economic growth. Our tribal nations, instead of selecting the most appropriate humans, tend to select the most powerful weapons. The sum of all these destabilizing positive feedbacks threatens Gaia herself.

As solar heat increases exponentially, so the speed of evolution has to increase in tandem if regulation is to be sustained. We can try sustainable development and renewable energy, and we can try geoengineering to help the Earth self-regulate. We can do these things with the same certainty that our eighteenth-century ancestors had about the power of mercury, arsenic or bloodletting to cure their diseases. Just as they failed utterly, so I think we also are not yet clever enough to handle this planet-sized problem and stop the Earth from overheating. But we can be fairly sure that if we do not survive, the lifespan of Gaia will be shorter than it could have been. Somehow we have to evolve to become this new part of the Earth system so that it can self-regulate better and cope with the dangers ahead. This is why I say that our survival is imperative.

But we have an enormous advantage, not possessed by those animals that evolved before we did. Because we have inventive ability and the science and engineering that allows us to explain and build, the natural selection of our mechanical artefacts has accelerated until it is now 1 million times faster than that of living organisms. This implies that potentially the lifespan of our now intelligent planet might be much longer than the 500 million years predicted by our rational Earth science that wholly ignored both our capacity to invent and the huge, 1-millionfold dilation of perceived time that comes with electronic instead of neuronic thinking.

Is it possible that our spontaneous move to live in cities could solve our climate and population growth problems as well? The termite nests with their air-conditioning towers that rise a metre or more above the desert are a wonderful example of the power of natural selection to optimize cooling, and the north–south orientation of the nests ensures that the removal of hot air from within the towers is maximized.

There are intriguing social possibilities if the ant or termite nest can be used as a model for human evolution in which we become a nest animal living in city nests. Would it bring a return to something similar to an idealized communist state, or a benign oligarchy – a state with a caste or class system and the disfavouring of democracy and egalitarianism? Invertebrate evolution in the nest reveals several possibilities: with termites the caste of workers are a dumbed-down version of the individuals from which they evolved. But if we use the bees' nest as a model we find the workers show remarkable individual intelligence for animals so small.

A great deal may depend on the extent of movement of people between the cities. If it is frequent and easy, the world may be relatively peaceful as it is now; but if the cities are isolated and travel

Figure 7. A termite nest.

between them is difficult, it seems likely that each city will form a tight tribal group and, as with social insects, wars between the city nests may be fairly frequent. Either way, life would continue much as we have known it for the past few millennia. The move to city living is more likely to improve than diminish our ability to use the Internet or whatever mass communication media replaces it. It will be good if virtual contact this way would suffice emotionally well enough to prevent tribal conflict. It would have an even better chance if the contact became more real than sound and vision alone. The popularity of today's social networks is encouraging, but they do little so far to calm the demons of tribal religious conflict.

The rational and decent concept of humanity as the ultimate unit for world governance seems so well established that it is pointless to challenge it. But what if in reality the tribe or nation is the largest possible unit? The existence of the United Nations seems to suggest a wish for something better than a reluctant peace between powerful nations and alliances of nations. Could it be more than this?

Natural history suggests that where Darwinian natural selection is the only force for innovation, evolution beyond the nests becomes very limited. In the past 100 million years social insects have persisted in much the same form as they first appeared. It seems possible that for humans the new, accelerated evolution will break this barrier and allow a more rapid evolution of their habitats.

In geological time, life can only continue on the progressively heating Earth if it evolves beyond our current form of wet carbon life. Ordinarily, to speculate on such a huge step would be pointless, but the rapid development of artificial intelligence might offer a lifeline from the future. Already we can see that an endosymbiotic fusion of neuronic and electronic life might be selected from its first primitive models. Now that the move to endosymbiosis might be under way, I can't help wondering how long it will be before an evolutionary step as great as that of the first eukaryotes changes the face of the Earth as much as they did.

It may be that we are now at one of those critical evolutionary bifurcations of our path. The future of the human species surely will depend to a great extent on how successful is the evolutionary acceleration path that we stumbled on 300 years ago. Whatever we do, if we succeed in establishing a new life era for the Earth, we will be remembered as the species that first harvested information and be as significant as that earlier species that first harvested the energy of sunlight.

10

So What Was It All About?

I feel truly blessed and privileged to have been raised in what was then the free and comfortable civilization of southern England. There was a sense of well-being and contentment that came from a belief that we were the world's leading tribe or nation. I am well aware that in 1919, the year of my birth, we had started down the long decline that still continues, but until our defeat by the Second World War it was easy to believe in the myth of Great Britain. So strong was this sense of privilege that, despite the great depression of the 1930s, I felt it as an adolescent in 1938 and used my pocket money to buy one of the first books of a collection that has never ceased growing. The book was *The Inequality of Man*, by J. B. S. Haldane, published by Penguin in their blue Pelican series of paperbacks. It was an extravagance, because before then the books I read were on loan from the public library. Haldane's book was first published in 1932 and its mild leftist views and strong, courageous support for science as a creative vocation, more than any other set the course of my life.

Was it pure chance that led this small island to be the birthplace of the Anthropocene? Perhaps it was, but familiarity with another island civilization, Japan, makes me wonder if long periods, more than hundreds of years, between wars, revolutions or natural disasters, are needed for a society to evolve beyond its turbulent childhood and provide an environment sufficiently stable for its Newcomens to flourish.

E. O. Wilson's wonderings about invertebrate sociality also bring in historical thoughts. Go to the nest of almost any social insect in the world and discover one of the most enduring aristocracies of all. It is extraordinary that even 100 million years of evolution has not displaced insect monarchs and their princes from their thrones. In societies that otherwise are models of egalitarianism, aristocracy endures. As the Victorians knew so well, and genetics now informs us, bloodstock and breeding are what really matter.

As if this fevered thinking was not enough, suddenly there came the concept of accelerated evolution. Like an effervescent tablet dropped into a brimming glass of strong liquor, it has fizzed in my mind throughout and shaped this book. Is it a real phenomenon?

Despite the decline of Britain's power, our ancestors would have thought that we now live in an enchanted world, and one that forever trespasses into the territory of magicians. It is as if we lived in a fairy story where magic carpets transported us effortlessly to whatever pleasure dome our hearts desired. Unlike our dreams or romantic tales, our waking lives are real enough, but somehow grow ever more difficult to distinguish from fiction. I find it hard to believe that as long as forty years ago I received a real message inviting me to fly to Florida to watch the launch of NASA's Viking mission to Mars, a mission that carried hardware I had invented and which helped to make its journey of discovery successful. Accelerated evolution allows us to take for granted the mechanisms of the car we drive, or the computer that takes our words and processes them. But with even greater disregard we still daily put to use the ancient but incredibly intricate mechanisms of our brains and nervous systems. Only since the invention of electronics and the science that followed have we been able to take the first few halting steps to understand what goes on behind our eyes. Then with deepening wonder we realize that it would be far easier to walk along the entire range of the Himalayas than understand the working of our brains; but they work so well that lack of comprehension rarely spoils the endless entertainment of consciousness as it proceeds and fills the spaces of our mind.

We suspect that we have little time left to deal with climate change, overpopulation, food and water shortage, and the other adverse consequences of our accelerated way of living. But how do we choose between the remedies on offer? Do we try sustainable development and renewable energy? Or do we bite the atom and rely on nuclear energy? Some offer to geoengineer the Earth to an ideal composition and climate. I think we might do worse than have trust in Gaia to regulate the Earth as she has done since life began, and retreat to the best cities that we can design and build with the objective of saving as many of us as we can; and entirely abandon the absurdly hubristic idea of saving the planet.

There is a larger environment, the solar system, that exists outside the confines of the atmosphere, and it affects Gaia much more than does humanity. The climate in the long term is dominated by the fact that stars like the Sun grow hotter as they age. The rate of increase of the Sun's heat may seem quite slow, but the ocean scientist Michael Whitfield and I argued in 1981 that the familiar natural world, the home of our forms of life, would be too hot in 100 million years. This is an unimaginably long time by human standards but not for Gaia, now about 3.5 billion years old. Our prognosis for Gaia was based on two facts: first, in the natural world the average temperature and carbon dioxide abundance in the atmosphere are closely connected; second, the output of heat from the Sun increases exponentially according to a robust and reliable understanding of the operation of its nuclear furnace. At present the flux of energy from the Sun shining on the Earth is 1.35 kW per square metre; in 100 million years' time it will have increased to 1.5 kW per square metre. That does not seem to be a large rise of heat input – little more than 10 per cent – but if the Earth system regulates its temperature by controlling the abundance of carbon dioxide then it would have to reduce it to zero to sustain present-day temperatures.

The Earth system can pump down CO_2 to relatively low levels, and reached 180 parts per million several times during the glaciations. But although plants need CO_2 to grow, evolutionary adaptations allow some of them to function at less than 100 parts per million, and at a higher average temperature than the current 16°C. As I mentioned in Chapter 5, Kasting and his colleagues at Penn State University suggested that by adaptations of this kind, life might survive as long as 500 million years. This is better, but still offers only a relatively brief old age for our planet.

But these are optimistic guesses: unfortunately, old planets, like old people, grow less resilient and increasingly vulnerable as they age, and cannot survive mishaps that they would have shrugged off when young. For planets, a mishap could be a planetesimal impact;

for people, a disease like pneumonia. Whatever caused the great extinction of the Permian period which ended 252 million years ago, it nearly extinguished Gaia. A repeat of this event now or in the future carries the probability of the end of life on Earth. Never forget that once extinguished, our carbon-based form of life could not start again at the Earth's position in the solar system.

The idea that one day, perhaps, the Earth itself may have a consciousness through the presence of our descendants has a long history. I first encountered it in the novels of Olaf Stapleton, in the 1930s. What is profoundly different now is the existence of practical hardware and systems that have evolved by and from intelligent artefacts. The great crowd of these individual intelligences will serve like the nerve cells of our own bodies. Among the many things they may do is sense the physical state of the environment and notice if it is too hot or too cold. Our collective minds will function together, in a way resembling the interaction of the neurons of your brain. Those neurons that are now active in reading this text are a tiny proportion of the total that go to make up your consciousness. The others, not asleep, are engaged in such vital tasks as checking and regulating your blood pressure, heartbeat, and the abundance of all the numerous chemicals that make up the normal composition of your blood. The neurons linked to the sense organs throughout your body all stay alert and are ready to warn of unexpected danger, or of pleasant things more worthwhile than continuing to read this book. All of them, except those asleep, are engaged in the myriad tasks of living, and are linked like a murmuring crowd. In the same way, my speculative vision is of a global consciousness composed of the billions of minds like yours and mine, either assisting or assisted by artefactual intelligence.

Already we are installing electronic artefacts within our bodies that communicate wirelessly with the technicians that control them. How long before our bodies are linked to the Internet? I must admit some angst at the thought that Service Pack 3 will be downloaded while I sleep.

On the much longer geophysical timescale of a billion or so years' time the Earth will approach a critical phase, one where liquid and solid water can no longer be part of the normal surface chemical mix.

Venus passed this phase eons ago and is in its long final chemical state, with an almost wholly carbon dioxide atmosphere and far too hot for any form of life of the kind we know. If we ignore the presence of Gaia, plain geophysical chemistry suggests that the Earth will be Venus-like in one or more billion years. By then evolution may have led to a clever system well able to adapt to greater heat or avoid it.

Long before the Earth becomes a rocky skeleton, it will be far too hot for the wet carbon-based organic life that we and nearly all life are made from. The upper limit for mainstream life is 50°C. It is true that highly specialized organisms can eke out an existence at temperatures at and even above 100°C in the hot springs near volcanoes. There have been many investigations of the upper temperature limit for 'extremophiles', and it is usually taken to be 120°C.

If we think in general scientific terms about life on a hot planet run by extremophiles, we see immediately that 120°C is much hotter than the boiling point of water at the present atmospheric pressure of 1.0 bar. Even 50°C is too hot by climate physics alone; it would imply a runaway greenhouse with water vapour as the principal greenhouse gas and a lifeless planet. Extremophiles have done well to adapt to life in boiling hot springs, but the laws of physics would need evading to allow them to form a biosphere. To expect extremophiles to regulate the climate of a planet running at much over 50°C is like expecting New York to be run by a mayor and his administration that were clinging precariously to the outer upper face of a skyscraper.

The real and present Gaia has been my mental companion for nearly fifty years, and naturally I am reluctant to see her eliminated by a form of cosmic global warming. But she is most unlikely to reciprocate these sentiments. Gaia's goal is to keep the Earth always habitable, but nothing in the rules say that carbon-based organic life is the only form allowed. Gaia has done such a wonderful job of keeping the Earth habitable for at least 3 billion years that it is not unreasonable to expect that it will continue to do so. But nothing suggests that our oxygen-rich world is permanent. The anaerobes and the prokaryotes gave way to us just as we and the other animals and plants may have to make room for another kind of world. Already we have evidence of the proto-life forms of such a new world, those that parasitize our computers; we even pay to have these viruses and Trojans destroyed.

Most biologists will assert that viruses, even organic ones like the polio or common cold viruses, are not alive because they cannot reproduce by themselves but have to borrow the bio-copiers of the real live cells they invade, just as computer viruses have to borrow the operating system of their host. Academic biologists who want to be accepted as scientists must cease claiming certainty about anything, especially life. I know that life is their subject, and they define it as something that can reproduce and then correct the errors of reproduction by natural selection. This is a fine definition, but I think it shows a lack of imagination to exclude Gaia merely because it does not reproduce in a familiar way. In the same way it is unimaginative to think that a virus cannot ever, by reverse engineering, evolve to become a living cell. The hardest part of life's invention must have been the composing of the symbols, the words, the sentences and the whole story written in nucleosides that forms the program that is life. There is no shortage of composed programs (sub-routines) in Gaia or in a virus. So rigid is the biologist's definition of life that if taken literally it implies that neither a post-menopausal woman nor a man sans prostate is alive.

What form of life could act physiologically as a regulator of climate and chemistry and keep the planet habitable when the Earth is too hot for the present organic carbon life form? Some science-fiction authors have toyed with ideas of other planets bearing life of a different chemical composition, usually silicon-based, but this was before the great revolution in understanding that came from the recognition that life's organizing and information-carrying principles were the aperiodic polymers named RNA and DNA. As a chemist, I feel sure that no element other than carbon could be the basis for such robust and vital information-carrying compounds as these, and given the mix of chemical elements that form planets, chemical life forms other than carbon-based seem wholly improbable anywhere in the universe. Not only this, but the spare parts from which carbon life originated are abundant in the gas and dust clouds of the galaxies. Consider the simple chemical, hydrogen cyanide (HCN), something once called prussic acid, which smells like bitter almonds and is poisonous. Infrared astronomy has shown HCN to be a frequent component of galactic gas clouds. By a few quite simple and probable steps, HCN can, with the assistance of light, water and free electrons, be the feed-stock for the nucleoside bases of DNA: thymine, guanine, cytosine and adenine. Other chemical components of life, including amino acids, have also been detected in space (see John Gribbin, *The Reason Why*, 2011). No comparable promising chemical feedstock lies around for a life form based, for example, on silicon chemistry. But as our computer viruses illustrate, life does not have to be based on cool wet chemistry. Schrödinger (see Chapter 3) defined life as an entity that takes in energy and raw materials from its environment and uses the instructions stored in its memory to make copies while excreting waste material and entropy to the environment.

We probably know what life is intuitively, but the devil is in the details. How does a tree basking in the sunlight use solar energy to split water molecules and excrete oxygen to the air, and at the same time take in CO_2 from the air and turn it into the solid substance of a tree? We now understand these details well enough and suspect that there is no clear-cut rational explanation that distinguishes self-organizing inorganic systems from life. In the mid-nineteenth century chemists spent much of their time falsifying vitalism, the theory that claimed that compounds of living things were unique. The debate closed when the chemist Friedrich Wohler showed that the inorganic chemical ammonium cyanate, when heated, became the unequivocally organic substance urea.

One thing I am sure about is that life cannot exist without a memory, and this must be a solid. How could the billions of bytes of ordered information in a molecule of DNA possibly carry its vital message as separate molecules of nucleoside dispersed in a gas or liquid? It would be like trying to read a book from a soup in which was dispersed all of the separate letters of the words of the text.

But what if there was a way to reboot Gaia with a new operating system on which to install its program, life? What if, as I touched on previously, the form of life changed from an organic wet chemical form limited to 50°C as the highest temperature for efficient metabo-lism, into a more physical life form, with an upper limit well above 100°C? Gaia's tenure might be extended even by as much as another 1 billion years. Such a physical system with electronic life could be based on any of these elements: silicon, germanium or carbon in an electrical form as graphene or diamond – or even simple compounds such as gallium arsenide – instead of a purely elemental form. There would still have to be a chemical process to extract and purify the raw materials of the Earth, and this could start from a symbiosis with organic life. Whatever its form, the new life would be subject to Darwin's law of natural selection, as were the species of its predecessors.

We must never forget that the priceless inheritance of humans includes the know-how of electronic hardware and intelligence. The new life, if its neurons operated at electronic speed and its design included intelligent software, could live 1 million times faster than we do and as a result its timescale would be increased as much as a millionfold. Time enough to evolve and diversify in the same way as carbon life has done. It might extend the tenure of Gaia still further, long enough even to enable the next Gaian dynasty, whatever that may be.

If anything as grand as this happened, the permutations and combinations made possible open up evolutionary possibilities as far beyond anything we can think of, far further than our thoughts now are beyond the comprehension of an intelligent Roman citizen. I see no point in speculating further except to ask that you keep in mind the fact that if such an electronic life emerged, its astronomical year might seem to it to last as long as 1 million years of our time, and the new Gaia would no longer be elderly but in the infancy of its lifespan. Time enough for some amazing natural and intelligent selection to happen.

Whether or not such intellectual brilliance emerges, we have to reckon with the huge cliff ahead of us in the Earth's journey into time. Our sibling planet, Venus, has been there before us and provides an awful warning of the ineluctable planetary death that lies in our own future. Venus started as an Earth-like planet, but being closer to the Sun evolved faster than the Earth. As the surface temperature of Venus approached the boiling point of water at the pressure of the Venus atmosphere, the sea, if it ever had one, would have boiled and in time the atmosphere would have become a mixture of supercritical superheated steam and carbon dioxide. Such an atmosphere is utterly inconsistent with our form of life, and probably with electronic life also. The supercritical state is real but we never encounter it naked. Carbon dioxide enters the supercritical state at temperatures above 31°C and pressures greater than 72.8 atmospheres; in this state it is neither a gas nor a liquid but has properties of both – it flows and mixes with other gases, but like a liquid it dissolves solids. If you are a decaf coffee drinker your coffee was probably rendered caffeine-free by washing the coffee powder in supercritical carbon dioxide. Water becomes supercritical at temperatures above 374°C and 218 atmospheres pressure, and in this state is so powerful a solvent that it will dissolve rocks. In the deeper chambers of the Earth gemstones, such as ruby and sapphire, condense from solution in critical-state steam, and the white veins of quartz we see in the rocks mark the passage of supercritical steam laden with silica.

Scientists interested in the inner planets of the solar system, especially Venus, Earth and Mars, assume that after the formative and violent period of the Hadean, the planets settled down with similar atmospheres and surface chemical compositions. Because of its position closer to the Sun it is difficult to envisage a scenario that allows Venus more than about 1 billion years before its oceans turned to steam and the surface rocks began to dissolve as the steam reached the supercritical state. The steam would have reacted with rocks like basalt and released the hydrogen of water as a gas. This would have slowly escaped to space from its upper atmosphere as hydrogen atoms, and in time have desiccated Venus entirely. The chemical reactions and heat would also have released carbon dioxide from any carbonate rocks, and this gas would be left to provide almost the whole atmosphere. This is what we see of Venus now, and the same disastrous progression to a hot and utterly inhospitable desert planet with a high-pressure carbon dioxide atmosphere is a near-certain fate also for the Earth. Fortunately it will be a long time before the Sun is hot enough for our oceans to become supercritical steam. Perhaps if our successors are intelligent enough, they could postpone such a fate almost until 5 billion years from now, when the solar system's inner planets are engulfed by the expanding Sun. As I have repeatedly said, Gaia's goal is to maintain a habitable planet but there is nothing in the rules that says the life form that inhabits it must be our present wet carbon-based form of life. There is a reasonable chance that the next form of life will be electronic and far less temperature-limited than we are.

It is a curious example of the malign effect of separating the disciplines of science that many astronomers still talk of a grossly swollen Sun, in perhaps 5 billion years from now, as setting the end of the familiar Earth we now enjoy. Much more probably, in less even than 1 billion years from now, a runaway greenhouse effect will turn the Earth into a planet as hot as Venus now and entirely lifeless. In a similar way, biologists who see 50°C as the upper limit for present-day life seem unaware that if the oceans were 50°C, water vapour would be present in the air at 8 per cent by volume, probably enough to start a runaway greenhouse effect.

Now is a critical moment in Gaia's history. It is a time of ending, but also a time of new beginnings. Despite the mess that we have made, carbon-based life still flourishes, and is likely to do so for tens of millions of years. This is more than long enough for electronic life to evolve and then either continue alone or in symbiosis with carbon life. Civilization may collapse, but there have to be enough humans, or intelligent successors, surviving to give Gaia the wisdom to proceed to the next step, whatever that may be, with or without us as the lead species. We can recover from a near extinction, as we proved one or two hundred thousand years ago, when the geneticists think that we were reduced to a few thousand breeding pairs. But now I see more clearly why we are crucially important for the survival of life on Earth. If we trash civilization by heedless auto-intoxication, global war or the wasteful dispersal of the Earth's chemical resources, it will grow progressively more difficult to begin again and reach the present level of knowledge. If we fail, or become extinct, there is probably not sufficient time for a successor animal to evolve intelligence at, or above, our level. It is the snail-like speed of natural selection and the slow transfer of information by wet chemical ionic conduction that hampers us and all other forms of animal life. Electronic conduction is 1 million times faster, but it needs our presence for the early and crucial stages of its emergence as a new life form. Our survival is therefore one of the most important steps in the evolution of our planet. Once another and faster life emerges and establishes a permanent base on Earth, Gaia might view the future with confidence.

Science fiction often plays out human dramas with a mixed cast of people and cybernauts and these can be fun, but there is a tendency to forget that large changes in the tenancy of Gaia can be slow. It took 1 billion years to move from the anoxic Archaean to the slightly oxygenated Proterozoic period and another 1.5 billion to reach the multicellular, oxygen-rich world of the past 700 million years. Even though electronic life may move 1 million times faster than we do, it still might find classical evolution unbearably slow.

It is easy to forget that modern science is born of the inflationary period, and does not follow seamlessly from classical scientific thinking. The expeditions to the upper atmosphere, to space, and the other planets, to the far reaches of the Earth, to the depth of its oceans, and down towards the magma, all have been made possible by the inflationary growth of engineering artefacts. The recent breathtaking evolution of computational equipment has made possible the most marvellously realistic models of the Earth, models so believable that we accept them as if they were the real Earth, and often fail to see that they are merely a simulation of it. The pioneers of computing, Babbage, Turing and von Neumann, were inventors and talked of their creations as machines. The rapid evolution of computers was driven by the needs of life-or-death cryptographic problems in the war. Turing and von Neumann were scientists as well as inventors, but as usual the need came first, then the inventions and lastly the science. The science then generated more needs, and so the positive feedback drove computer speed and capacity to double its rate about every two years, enough to increase the speed and memory of computers 40 billion times since the cryptographic computers of Bletchley Park in the 1940s.

Vinge and Kurzweil have speculated that in a few decades artificial intelligence could equal human intelligence and then proceed to double every two years, if Moore's law continues to hold true. Most of us feel instinctively that these 'singularity' nirvanas or dooms are unlikely to happen, and even if they did, the machines would still be our slaves and easily be disciplined to continue to serve us as before. Even if there were a cyber revolution, we suspect that Spartacus, the steward of the new revolution, would as usual fail when confronted by a virus, treason, or even its Service Pack 1. But we would be very stupid if we forgot that Spartacus might be able to think 1 million times faster than we do, and might even use intuition.

If we are really concerned about the threat from an artificial intelligence greater than ours, I think we should immediately start a programme to discover more about the brains of the great whales. Their brains are larger and richer in neurons than ours. Do they have intelligence proportional to the size of their brains? We had better find out before population pressure among humans drives us to eat them all.

The overwhelming competitiveness of evolution by natural selection must play a huge role in the next steps, not least in stimulating the aggression which we would feel towards machines that we feared were more intelligent than ourselves. But what is intelligence in this context? So far we have not yet devised a convincing way of measuring or comparing intelligence among humans. The crux may depend on a full definition and comprehension of what intelligence is, and perhaps then the realization that understanding intelligence is far beyond our capacity. The intelligence test measures only a limited part of rational intelligence, which may be why a high score does not seem to correlate well with achievement, and may explain why membership of Mensa is not more prestigious.

J. B. S. Haldane said long ago, 'My suspicion is that the Universe is not only queerer than we suppose, but queerer than we can suppose', and Vlatko Vedral in his recent book (better than any other I have read on the subject) *Decoding Reality* (2011) took me through the first steps towards the meaning of reality and helped me to accept how little of it I will ever understand. Modern physicists who have thought hard and long about the nature of the universe seem to say that everything in the end is information – another word, like intelligence, which we think we can define but which so far eludes our conscious minds.

What I do understand, and have tried to convey in this book, is that rational thinking is not necessarily our greatest property, and although we prize it, it may be a handicap. We have to recognize that in addition to conscious rational thinking our minds are capable of other, more powerful mental processes that lead us by intuition to grasp a tiny sparkling fragment of reality.

Intuition must be taken into account when comparing the power of artificial intelligence with animal intelligence. Does any current or forthcoming computer have the capacity for intuition? Then there is that awesome thought: what if our brains are quantum computers? And if they are, does this have anything to do with our ability to think using the hidden layer of the mind that operates unconsciously?

Obviously, artificial intelligence has already surpassed us by astronomical orders of magnitude in some of the things it can do. Why, even the PC on my desk, vintage 2012, can perform feats of arithmetic that I could never ever do, such as expressing the largest prime number it could calculate in a second. But we should not be overawed by this. It may not be more significant than the fact that your car moves much faster than you could ever run.

Scientists are not all cold-hearted like Mr Spock, and many of us like to imagine the far future. For the astronomers among us, the recondite pastime is speculating on the past and future of the universe. What seems more interesting to us now is the detail of the near future on Earth. Just because it is possible to imagine all kinds of electronic life forms, their molecule-sized nuts and bolts, and the algorithms needed for autarky, this does not mean that such constructs can easily be made or that they represent the life forms of the near future. To imagine that carbon life has exhausted all of its possibilities and has no future in a warming world is massively wrong, especially now that we have started to interfere by modifying genes. Darwin himself was inspired by the powerful capacity of selective breeding as long as 160 years ago. It is scary to think how fast new species can be made by direct molecular intervention. It might lead to the establishment of a world run by genetically modified extremophiles. If this is possible it has the potential to extend considerably the wet carbon form of Gaia's lifespan both alone and as endosymbionts with electronic life.

Then there is the transition of Gaia from carbon-based wet chemistry with ionic conduction (our way of handling signals in the nervous system) to a future symbiosis with electronic life. But what if we have already taken the first steps? Lynn Margulis envisaged the phagocytosis of a cyanobacterium by a larger bacterial cell, with the occasional consequence that the two organisms lived in symbiosis for their mutual benefit. What made her insight remarkable was that it explained the emergence of the first plant cell from which all vegetable life has evolved. She called this crucial new combination endosymbiosis, and it also explains the emergence of muscle cells and neurons in animal life.

If we can somehow merge with our electronic creations in a larger-scale endosymbiosis, it may provide a better next step in the evolution of humanity and Gaia. Fictional tales of monstrous battles with cybernauts, always with us as the winners, are unconvincing; more likely would be our union with electronic life, which has already taken its first tentative steps. All things that evolve do so in a series of small but finite steps, and the first of these towards electronic symbiosis was taken in 1958 when the Swedish surgeon Ake Senning implanted a cardiac pacemaker designed by Rune Elmqvist into Mr Arne Larsson, and it worked. The next steps have already been successfully achieved, and the implanting of computer-controlled pacemakers for human hearts is now a routine procedure. My personal pacemaker already has a transmitter and receiver so that a cardiology technician can assess its performance and make adjustments, if they are needed, without further surgery.

Having done this for the heart and for several other similar simple repair systems for ailing or ageing bodies, such as knee and hip joint repairs and electronic implants to remedy sensory defects in vision and sound, we would still be a long way from a separate self-powered and self-sustaining union of carbon and silicon life, which when completed might be something similar to but not the same as the cybernauts or androids of science fiction. The 'bionic' people made this way could not at first reproduce sexually. Not even if the implanted equipment was evolved from the human body by genetic means. It would require a great deal more tinkering with the DNA of eggs and sperm to build a new form of life that bred true.

Now that my body carries the means of a two-way connection, I must admit an empathetic dread for some unfortunate future person whose body becomes connected to one or more of the ubiquitous social networks. I suffer with that future person and can imagine no punishment more severe than having my still comparatively clear mind overtaken by the spam of hucksters and the never-ceasing gossip of the Internet.

Proud and talented as we are, it would be quite a step to make an entirely novel life form. I am as near certain as a man of science is allowed to be that we could never take even the first steps of so grand a project consciously and rationally. First, we have to be sure that we need these new life forms. Remember that need is what inspires and empowers our minds to invent, and is the driving force of evolutionary inflation.

Those are some of my thoughts on a few of the personal and Gaian consequences of evolutionary inflation. What about the consequences for humanity? First we have to ask, would there have been any of the present-day pollutions, climate change and population growth if the inflation had not happened? Let's assume that after 1700 there was no attempt to make a steam engine do work and no surge of inventions happened. Let us also assume as a thought experiment that evolutionary inflation did not start until later in the twenty-first century.

Without accelerated evolution, would the sum total of our needs have changed much from those of previous centuries? We would have still needed cures for many diseases, including malaria, plague, leprosy, influenza, typhoid, typhus, measles and smallpox. The discovery of vaccination by Jenner might still have happened, and other low-tech cures and inventions have been made. Soon the invention of microscopes powerful enough to see bacteria clearly would have established the bacterial origin of some disease, and in time more cures have been invented. The outcome of these inventions would be a growing pressure for more energy, and the steam engine would have appeared not much later than it did. We are unlikely to have remained stalled in a preindustrial steady state, although we might wonder why the Chinese did.

The inventions of the Anthropocene raised the standard of living of ordinary people so much that a culture of preventative medicine, hygiene and waste disposal became possible in the wealthy nations. This is something that works better than antibiotics in the cure of disease. There is nothing fanciful about this statement. I have frequently been surprised during my lifetime at how few physicians and medical scientists seem aware that many previously deadly infectious diseases including diphtheria, haemolytic streptococcal infections, syphilis and tuberculosis declined in severity at much the same rate before and after the use of antibiotics. The improvements in living conditions and the gift of evolutionary inflation were at least as important as specific cures. The association of health with a culture of preventative medicine is confirmed by the high incidence of disease today among migrants to the first world who accept the benefits of welfare but retain their original cultures.

Progress has carried us a long way but at last shows signs of slowing. While it continued its rapid motion there was not much chance for us to do anything other than watch the scenery as it passed and hoped our destination was a pleasant one. This I think is the main reason for our inaction on climate change; we all went to Kyoto full of good intentions and out to save our world but it was moving so rapidly there was little that we could do.

In Chapter 7 I discussed the possibility that we might resolve the problem of global warming for most of us by retiring to live in giant cities. A mainly tropical world at 20°C offers a climate cooler than Singapore today; perhaps we should accept that both human nature and Gaia will thwart our efforts to restore the status quo, and instead relax in what the travel agents offer as a tropical paradise. Of course there is much more to climate change than global average temperature. A rise of 5°C in the global average might bring huge changes in the distribution of temperatures, and in the quantity of rain and snowfall. The sea level would rise and there would be changes in the distribution of land, ocean, deserts and forests. I still think that well-chosen city sites would offer us a better chance of survival.

I am, as I hope you will have gathered by now, an optimist. I do not envisage the death of Gaia, the Earth system, in the immediate future, either through human folly or otherwise. It can sustain human life for a good while yet, and human life can be the catalyst for Gaian survival in the much longer term. But there is one snag. The system cannot sustain the present level of human population for very much longer. The future world may be a better place, but getting to there from here will not be easy, and we will not all make the journey.

Appendix

TRANSCRIPTIONS OF LETTERS FROM W. D. HAMILTON AND J. MAYNARD SMITH

(see pp. 80–81)

First page of letter from W. D. Hamilton

3 March 1995
Dear Jim,

I had not read far enough in your paper when I wrote yesterday to realize that the degree to which a daisy reflects affects its own temperature and thereby its own growth. Thus I wrote as if I believed the selection on a particular daisy from having a given albedo was extremely small or zero. I see now that this is not the case and that there can be strong differential survival depending on albedos. The reflective property of a given daisy becomes therefore strongly important for itself and with side effects for the environment that happen to be stabilizing. If I had realized this our discussion all along would have been rather different and it was certainly foolish of me not to look at least as far as the next page and notice that you had a brief discussion of 'cheating' there already! I therefore have to re-aim a large part of my criticism at your assumption that what is good for the individual is likely to be stabilizing for the environment. The case of my benthic microbe of course is just the opposite: it is advantageous for the individual microbe to get rid of its waste product but this happens to be highly destabilizing for the world as a whole – and the benthic organism is the last to notice the effects of this.

I can see now that with the assumptions of the daisy world model it is not going to be possible to cheat by discarding the lamina because the individual needs one to regulate its own temperature. In fact the 'cheat' of your Figure A is really just one of the sequence of daisy types that you display in Figures B and C except that it has a 5 per cent advantage, so that if present in the other diagrams it would produce a tiny blip (about 5 per cent up I suppose) in the smooth progression of the other types, and a blip centred at about 0.85 solar luminosity. The fact that you get just two daisy types under prolonged selection at a given solar luminance is just what I would expect, and I think I now see how it is that Stephan's herbivores, if they tend to concentrate on whatever is the most common will prevent this happening, and thus, through keeping more daisy types around, improve the system's ability to adjust rapidly if Solar luminance is varying back and forth also rapidly (I vaguely remember something like this being in the paper).

First page of letter from John Maynard Smith

25 March 1993
Dear Jim,

Forgive me for the delay in answering your letter – I've been travelling around.

First, the easy part. I'll try to keep April 4–8 1988 free. I wouldn't want to present a formal paper, but I'd like to come.

Then, the hard part. I think you are right in seeing a hostility to Gaia among professional population biologists – particularly neo-Darwinists. I understand the reason. The reasons are not excuses. What neo-Darwinists have been trying to do is to apply notions of natural selection in a rigorous way. They saw the Gaia hypothesis as a loose and unjustifiable extension of evolutionary thinking. But why should this have aroused hostility? After all, one can disagree with someone without being hostile. I think the reason is that Darwinists, too, have felt themselves to be a persecuted group. This will sound ridiculous to you, since Darwinism is the scientific orthodoxy. However, although a scientific orthodoxy, it has been widely rejected by a wider educated public. People just don't want to believe that they are the products of random mutation and selection. They want to believe that God loves them!

Further Reading

BOOKS ABOUT THE SCIENTIFIC INFRASTRUCTURE

Martin Rees (2003), *Our Final Century*
Vlatko Vedral (2011), *Decoding Reality*
J. B. S. Haldane (1927), *Possible Worlds*
Erwin Schrödinger (1944), *What Is Life?*
Sir John Houghton (2009), *Global Warming*
Kendall McGuffie and Ann Henderson-Sellers (2009), *A Climate Modelling Primer*

BOOKS ABOUT EARTH SCIENCE

Michael C. Jacobson, Robert J. Charlson, Henning Rhode and Gordon H. Orians (2000), *Earth System Science*
Lee R. Kump, James F. Kasting and Robert G. Crane (2004), *The Earth System*
David M. Wilkinson (2006), *Fundamental Processes in Ecology*
Tim Flannery (2011), *Here on Earth*
Stephan Harding (2009), *Animate Earth*
Tim Lenton and Andrew Watson (2011), *The Revolutions That Made the Earth*
E. G. Nisbet (1987), *The Young Earth*
Oliver Morton (2007), *Eating the Sun*

BOOKS ON CLIMATE CHANGE

Stewart Brand (2009), *Whole Earth Discipline*
Burton Richter (2010), *Beyond Smoke and Mirrors*
James Lovelock (2009), *The Vanishing Face of Gaia*

BOOKS ON BIOLOGY

Bert Holldobler and E. O. Wilson (2009), *The Super Organism*
W. D. Hamilton (1996), *Narrow Roads of Gene Land*
E. O. Wilson (2012), *The Social Conquest of Earth*

BOOKS ABOUT THINKING BY HUMANS, ANIMALS AND MACHINES

John Gray (2013), *The Silence of Animals*
Michael Shermer (2012), *The Believing Brain*
Donald R. Griffin (1992), *Animal Minds*
Daniel C. Dennett (2013), *Intuition Pumps and Other Tools for Thinking*
Ray Kurzweil (2012), *How to Create a Mind*
Ray Kurzweil (1990), *The Age of Intelligent Machines*
Daniel Hillis (1998), *The Pattern on the Stone*
Marvin Minsky (2006), *The Emotion Machine*

Index